SPECTACLE EARTH

Cultural Frames, Framing Culture

Robert Newman, Editor
Justin Neuman, Associate Editor

SPECTACLE EARTH

Media for Planetary Change

Andrew Kalaidjian

University of Virginia Press • *Charlottesville and London*

The University of Virginia Press is situated on the traditional lands of the Monacan Nation, and the Commonwealth of Virginia was and is home to many other Indigenous people. We pay our respect to all of them, past and present. We also honor the enslaved African and African American people who built the University of Virginia, and we recognize their descendants. We commit to fostering voices from these communities through our publications and to deepening our collective understanding of their histories and contributions.

University of Virginia Press

Printed in the United States of America on acid-free paper

First published 2025

9 8 7 6 5 4 3 2 1

ISBN 978-0-8139-5271-0 (hardback)
ISBN 978-0-8139-5272-7 (paperback)
ISBN 978-0-8139-5273-4 (ebook)

Library of Congress Cataloging-in-Publication Data is available for this title.

Cover art: Augsburg Book of Miracles, folio 144, 1542
Cover design: Trudi Gershenov

For Elizabeth

CONTENTS

List of Illustrations ix

Acknowledgments xi

Introduction: Mediacene 1

Part I. Golden Spikes and Splices

1 Aesthetics of Antimediation 21

2 Natural Historicities 46

Part II. Whole Earth Resistance

3 From Global Village to Anthropocene 73

4 From Ecological *Détournement* to Rebel Geology 97

Part III. The Golden Age of (Dis)content

5 Cyber Saturnalia 121

6 Digital Twins 142

Conclusion: Reality's Return 165

Notes 171

Bibliography 187

Index 197

ILLUSTRATIONS

1. *Cardinal Albrecht of Brandenburg as Saint Jerome,* Lucas Cranach the Elder, 1526 27
2. *Lion,* Louis Rhead, 1807 35
3. *The Shipwreck,* Benjamin Smith, 1797 41
4. *Weather Forecasting Factory,* L. F. Richardson, 1922 44
5. *Waterloo Bridge,* Claude Monet, 1903 51
6. Imperial Institute, exhibition galleries and cinema sign 61
7. Imperial Institute, life in the colonial empire 62
8. *Newton,* William Blake, 1805 79
9. *Narciso,* Caravaggio, 1594 88
10. *Dragged Mass,* Michael Heizer, 1971 92
11. Untitled, Constant, from *Labyrismen,* 1968 115
12. *Saturn Devouring His Son,* Francisco Goya, 1823 131
13. *Seedlings_:From Humus,* Qianxun Chen and Mariana Roa Oliva, 2021 139
14. *Garden,* Owen Roberts, 2021 140
15. Illustration for "Silence Is a Commons," Matthew Wuerker, 1983 147
16. *Dump VR,* Vojtech Leischner and Natalia Peterkova, 2019 153
17. *For Submersion,* Sarah Rosalena, 2023 157
18. *Mushroom Cloud,* Nancy Baker Cahill, 2022 162

ACKNOWLEDGMENTS

This book builds on ideas originally published as "The Spectacular Anthropocene" in the journal *Angelaki* in 2017. Parts of chapter 2 appeared in different form as "Historicities" in the *Oxford Encyclopedia of Literary Theory*. Thank you to colleagues in the Society for Literature, Science, and the Arts, the Electronic Literature Organization, the Modernist Studies Association, the British Association for Modernist Studies, and the Association for the Study of Literature and Environment for helping to develop those initial pages.

Research support from California State University, Dominguez Hills College of Arts and Humanities, the CSU Chancellor's office, the Grants for My Research program, as well as the Sutton Award for Research in English provided release time to complete much of the writing. I'd also like to thank CSUDH graduate and undergraduate students in my courses on Lit/Media/Tech and Climate Fictions for engaging with much of this material and providing insight into mediated life.

The University of Virginia Press, editor Angie Hogan, and the reviewers of the manuscript provided incredible feedback that led to two substantial revisions of the project. Special thanks to Joe Figliulo-Rosswurm and Paul Megna for providing feedback on earlier versions of the manuscript.

My appreciation goes out to the libraries, museums, and galleries that helped with archival research and image licensing, including the National Archives, Stanford University Special Collections, the Museum of Modern Art, the Guggenheim, Tate Britain, Denver Art Museum, Museo Nacional del Prado, Gallerie Nazionali Barberini Corsini, Gagosian Gallery, Folger Shakespeare Library, Library of Congress, the Cartin Collection, and the Ringling Museum.

Thank you, finally, to all of the artists who provided images for inclusion in the manuscript and gave generously of their time to discuss their work including Michael Heizer, Owen Roberts, Qianxun Chen, Mariana Roa Oliva, Sarah Rosalena, Christian Nyampeta, Nancy Baker Cahill, Vojtech Leischner, Natalia Peterkova, Matt Wuerker, and Stephen Conlin.

SPECTACLE EARTH

INTRODUCTION

Mediacene

> The spectacle is the sun that never sets on the empire of modern passivity. It covers the entire globe basking in the perpetual warmth of its own glory.
>
> —Guy Debord, *Society of the Spectacle*

> The future isn't what it used to be.
>
> —Tracy K. Smith, *Life on Mars*

A RECENT strand of environmental activism has taken aim at the arts. From throwing soup on Da Vinci and Van Gogh to interrupting Broadway shows, groups such as Just Stop Oil and Extinction Rebellion are finding new ways to keep planet Earth in the headlines. Unfortunately, the more nuanced discussions behind these actions about food sustainability, political engagement, and environmental justice are often lost in the ensuing backlash. One study even suggests that such actions have a detrimental effect on public support for efforts to address climate change.[1]

Behind this public outcry might also be a sense that the activists are engaging in a bit of friendly fire: Aren't the arts on the side of good? Doesn't the world need more sensitivity to the human condition, greater respect for the works of past generations, and new forms of representation in artistic mediums? If these activist efforts meet with mixed results, they are more successful in bringing attention to the long and complicated relationship between environmentalism and spectacle.

These strange acts spring from even stranger changes to the planet. From devastating floods to record forest fires, heat domes that cook shellfish alive, ice storms in Texas, and tropical temperatures in Siberia, unprecedented events are affecting life around the world. While the more catastrophic effects of climate change make headlines and the evening news, advanced media technologies provide new insights into the longer-term repercussions of life in the Anthropocene. Time-lapse videos show the melting of glaciers and the bleaching of coral reefs. Earthrise Media's

Climate TRACE deploys artificial intelligence and machine learning to visualize real-time greenhouse gas emissions throughout the world. Donning a virtual reality headset allows users to interact with digital twins of endangered species and threatened habitats. The Green Game Jam encourages video game developers to implement green activations that empower gamers to play for the planet. Your favorite band has installed kinetic dancefloors so that your moves can offset carbon on their world tour. Thanks to television, smart phones, the internet, satellites, computer modeling, and a slew of other technologies, today there is greater environmental awareness than ever before.

The changes can be seen, experienced, and felt in novel ways, but the question remains: Do humans recognize their own activity behind Earth's degradation? Will media saturation of all things ecological lead to a more sustainable future? Or, to borrow from Paul Virilio, will the planet's demise merely become "one series of scopes among others"?[2] The dilemma is a loop between raising awareness and becoming inured to ongoing crisis. Mediums that highlight environmental problems can also hinder meaningful engagement with these same issues. If not met with direct action, an abundance of ecomedia risks becoming just another stream of content.

Scholars throughout the environmental humanities have begun to address these concerns. Several works call attention to the physical impact of media: the toll taken through raw materials, rare mineral extraction, growing e-waste, and carbon emissions. Others curate examples of green media: art, literature, music, film, video games, and digital projects that are helping to represent environmental issues and realign human perception to act differently toward the nonhuman world. What remains less explored, however, is the extent to which media technologies normalize anthropogenic change.

A life increasingly spent on screens and devices leads to the distraction, passivity, and shallow commitment necessary to sustain the Anthropocene as a geological reality. Rather than outlining a more perfect ecological mediation, this book explores cultural histories of planetary media and theories of *antimediation* that consider the benefits of disconnection. But it is not as simple as turning off the television and walking away. If that were the case, there would be no need to write or read this book at all. Instead, the challenge is to create disruption within media: to bring attention to the mediums, the politics, and the histories of creative environments. Moments when media reveal the structures behind their messages can unsettle a person's frame of reference, opening the door to a different relationship with the world.

Nature, as John Durham Peters reminds us, is the great mediator through the classical elements of air, water, fire, and earth.[3] Calls to geoengineer the planet indulge in the fantasy of controlling these elements, just as Shakespeare's Prospero controls the tempest. But nature is also the great antimediator—changed, augmented, and *détourned* through human expenditure—it wastes civilization's best laid plans with flood, drought, forest fires, earthquakes, and hurricanes. These world-changing events become media focal points of spectacle. But beyond the scramble to restore the power grid and rebuild the levee, the overwhelming disruption of disaster is an opportunity to rethink forms of connection in an increasingly networked world: an epoch that environmental media scholars have begun to theorize as a Mediocene, AnthropoScene, or Media(s)cene.

Life in a Mediacene

It is important to ask who has access to sophisticated media technologies. What interests seek to control and shape the media landscape, and how can efforts to democratize accessibility to these tools avoid further extending their undesirable effects? In 2018, the Internationale Kolleg für Kulturtechnikforschung und Medienphilosophie (IKKM) held a conference on the "Mediocene" and subsequently published a collection of essays that explore media in the human epoch. The term "medio" is a corrective to "anthropos" that calls attention to a larger process of mediation driving planetary change. At its broadest, the "medial" is yet another term for environment, surroundings, network, or any intermediary relationship between subject and object, organism and ecosystem, individual and community. The medio also signals the technical layer that humans have developed over time to shape their interactions with the world.

While this focus works against the narrative of humans as the controllers of global destiny, Georg Toepfer points to a peculiar paradox that comes from foregrounding media processes: "Media, as devices that connect things, tend to eliminate the environment as something which is outside of the system. . . . The mediocene as the era of the environment, or of environmental concern, is being eradicated by the mediocene as the era of media processes."[4] In other words, as the media environment grows, synthesizes, and integrates lived experience, it draws attention away from the physical world, all while increasing the extraction and consumption of natural resources. The following chapters explore this paradox through a focus on comparative media.

Literary and artistic works have long engaged with the rising tide of media influence in modern society. Don DeLillo's 1985 novel *White Noise* presents Jack Gladney and his family, who can't get enough of environmental spectacle but are unprepared when actual disaster arrives on their doorstep. The viewing of dramatic events takes on the status of ritual: "That night, a Friday, we gathered in front of the set, as was the custom and the rule, with take-out Chinese. There were floods, earthquakes, mud slides, erupting volcanoes. . . . Every disaster made us wish for something bigger, grander, more sweeping."[5] Later, in a conversation with colleagues, Jack wonders why he and his family were so drawn to the screen. The answers he receives note the importance of the catastrophic for breaking up the day-to-day "brain fade" as well as simple schadenfreude. The perverse desire to view larger and more spectacular disasters arises in relation to the mundane flattening of information in the media age, where an advertisement for dish soap exists on the same plane as a forest fire. Further, the world as it appears on TV is not given second stature to lived experience: "For most people there are only two places in the world. Where they live and their TV set. If a thing happens on television, we have every right to find it fascinating, whatever it is."[6] The Gladneys are not to be judged for their engagement with media: it is universal, natural, human.

Since the 1980s, the second world of media has grown from the familial television set to an interconnected realm of devices and content. Beyond sensationalist news coverage and Hollywood disaster films, the twenty-first century has become a golden age for environmental media. Al Gore's 2006 documentary *An Inconvenient Truth* launched a wave of ecocritical films, and the BBC's 2006 production *Planet Earth,* the first nature documentary filmed in high definition, has led to countless hours of vibrant, immersive wildlife content. Studies question the long-term efficacy of the dramatic "wake-up call" tone often taken in environmentalist films and TV.[7] At the same time, seeing nature and wildlife on screens can lead to increased happiness and well-being.[8] People gain the luxury of connecting to nature from the comfort of their homes while habitats disappear.

If the spectacular nature of the Mediacene creates new opportunities for the representation of ecological issues, an increasing reliance on technology can also isolate individuals, polarize communities, discourage activism, and foreclose avenues to create meaningful environmental change. In this sense, a digital media epoch lives up to Guy Debord's theory of the spectacle as "separation perfected" that opens his 1967 work, *The Society of the Spectacle.* As Jonathan Crary notes, spectacle for Debord goes beyond a visual phenomenon of entertainment: his real interest lies in a system of parameters—often unseen—that individualize, fragment, and detach people from each other.[9] Spectacle is an architecture imbricated with the "non-places" that Marc Augé identifies as avenues of isolation. Augé considers how media immersion can become its own non-place, allowing people to live in an "intellectual, musical, or visual environment that is wholly independent of immediate physical surroundings."[10] Crary goes further to explain how capitalist systems condition humans to actively cancel out their environments to better pay attention to specific tasks.

What is necessary to disrupt the placid consumption of media and motivate people toward real change? In his epilogue to *Non-Places,* Augé considers the classic sci-fi trope of the discovery of alien intelligence and its resulting effect of ending intraspecies disputes. This is the setup of numerous books, TV shows, and movies such as *Ender's Game, Star Trek,* and *Independence Day.* In these scenarios, humans either put aside their differences and join in a greater intergalactic community or, more often, assemble their combined military forces to obliterate the unwanted incursion.

As aliens continue to shun Homo sapiens in the twenty-first century, the next best candidates for species unification are large-scale environmental transformations such as climate change, resource scarcity, industrial toxicity, ocean acidification, and species extinctions. Augé, for his part, is skeptical, noting that when it comes to building solidarity and seeing the planet as a single place, "it is far from certain that threats to the environment are sufficient."[11] The introduction of the term "Anthropocene" to denote a geological epoch driven by human activity is just such an attempt to *universalize* the environmental issues facing humanity.[12] Critics ask what exactly defines the "human" in these scientific debates, given that some people and ways of life have a more detrimental effect on the environment than others.[13] Environmental humanities scholars note the issues of intangibility and alienation that accompany thinking on a global scale.[14] Others pose the work of ecologically minded artists and writers as crucial to help the public become "affectively engaged" with environmental issues.[15]

Media theorists have jumped into the fray as well to question the role of images in environmental discourse and to call attention to the mining of rare minerals and the exploitative labor conditions that supply the ever-increasing demand for devices. Stacey Balkan claims that "environmental violence is notoriously spectacle-deficient" while "the populace is routinely paralyzed by the latest cli-fi blockbuster—terror quickly transformed into inertia."[16] For those with the privilege to view human suffering from a distance, Anders Ekström identifies a new "ordinariness" of disasters that have been "literally miniaturized and no longer defined by the sublime."[17] T. J. Demos poses the aesthetic question succinctly: "How does the Anthropocene enter into visuality, and what are its politics of representation?"[18]

These critics often situate the flurry of attention given to environmental crises alongside Rob Nixon's discussion of "slow violence": the host of toxic and accretive environmental issues that have the potential to cause exponential harm over time. Unlike traditional violence, which is highly visible, newsworthy, and clearly defined in terms of perpetrators and victims, slow violence is often unseen, dispersed, and difficult to prove in a court of law. Nixon aligns the rise of anthropogenic change and the growth of the internet, arguing that "high-speed planetary modification has been accompanied . . . by a digital world that threatens to 'infowhelm' us into a state of perpetual distraction."[19] He grounds his ensuing approach to environmental justice in a critical discussion of representation, visibility, and narrative.

While aesthetics can galvanize activism, it is important to interrogate mediation itself to better understand the potentials and pitfalls of addressing environmental issues through indirect channels. The recent "greening" of media content does not necessarily work against the subsumption of environment and, indeed, may at times be complicit with this process. Nor does marveling at the vastness of media infrastructure provide an easy escape from reliance on these systems. Environmental media studies prove very successful at analyzing ecological themes and representations in art, film, television, literature, as well as newer digital forms, video games, the internet, and virtual reality. Yet the bridge between this ecocritical approach to media and tangible environmentalist reform remains tenuous.

While this book is not primarily focused on activist movements, it is interested in resisting forces that work against political engagement: Specifically, how does the growing media realm foreclose the necessary attention and perseverance needed to create meaningful change? Caroline Levine's recent work *The Activist Humanist* calls attention to the challenges facing environmental humanities scholars seeking to bridge the gap between ecocriticism and activism. Levine convincingly faults the aesthetic humanities for promoting anti-instrumental critiques of open-endedness, potential, and rupture without adequately developing a practical way forward for meaningful changes to take place. She turns instead to Kyle Powys Whyte's concept of "collective continuance" to outline a formalist process of small gains that will have a growing influence over time.[20]

Yet any appeal to formalism in the twenty-first century should consider the ramifications of form's erasure of matter through technical media. Friedrich Kittler points out that as early as 1859, Oliver Wendell Holmes's essay "The Stereoscope and the Stereograph" ends by prophesizing the divorce of form from matter, with the result that once technology harvests the "skins" of nature and great human edifices, there will be little incentive to keep or maintain the physical versions.[21] What has followed has been a bit more complicated. The coliseum has not been torn down despite the millions of photographs that exist of it. In fact, the physical real has become more valuable, as people desire to go to the source, to capture a skin of their own. While historic sights and sublime environments have no doubt benefited from a proliferation of media representation—Ansel Adams's Yosemite photos, for example—Holmes is on to something in linking a growing disregard for the material world to the emergence of a virtual one.

At the same time that technical media assimilates and synthesizes matter and artistic techniques, a new rhetoric of "media ecology" emerges to welcome this process. As Matthew Fuller explains, "Such environmentalism also often suggests that there has passed, or that there will be reached, a state of equilibrium: that there is a resilience and harmonic balance to be achieved with some ingenious and beneficent mix of media."[22] Fuller and others challenge this idyllic view of ecology through the injection of postmodernist disequilibrium, asymmetrical relations, and messy assemblages.

While many environmental media scholars take comfort in elaborate theories of mediation that may be able to parse the complex interactions between technology, nature, and culture, recent work by Anna Kornbluh suggests a new challenge that arises as traditional forms of mediation give way to a new preference for immediacy, a style that promises direct access and connection. Instead of a revolutionary aesthetic, however, Kornbluh convincingly shows how supposedly unmediated forms are often symptomatic of resignation to and complicity with a planetary capitalist system that is too far gone.[23] Just as the Mediacene is named and theorized, it is already giving way to the Immediacene.

Maxims for Resisting Mediation

In 1930, George Halcott Chadwick proposed the term "Phanerozoic" for the current geological eon. Combining the Greek *phanerós* (visible) and *zoē* (life), he argues that "life is after all the main criterion."[24] For the last half-billion years, complex flora and fauna have appeared and disappeared, evolved and gone extinct. Now, one species, Homo sapiens, is leaving an outsized imprint on the history of Earth. In seeking to legitimize the human epoch in the eyes of the scientific community, geologists present their work as a measurement of reality: one that they pursue to a high degree of certainty using stratigraphic analysis. But too often data collection becomes its own end. Obsessive observation may foreclose the necessary intervention that would mitigate disaster. Don't worry! the data exclaim seductively; we are monitoring the situation—we have an eye on it!

Let the first maxim for resisting the mediation of ecological crisis be clear: *observation ≠ action.* The notorious and deadly London fog was one of the most technically and obsessively monitored environmental threats of the twentieth century. Beginning in 1912, scientists of the Committee for the Investigation of Atmospheric Pollution developed new technologies to detect smog such as air filters to collect soot and particulates. Visual gradients and photography were used to compare the appearance of smog to measurements of its toxicity. Forty years of data did little, however, to avoid or alleviate the catastrophic Great Smog of 1952, which descended on the city and resulted in over four thousand deaths. In becoming a sensation, a disaster, a spectacle, the Great Smog did finally provide the impetus to make meaningful reforms in the Clean Air Act of 1956 that would lead to a 60 percent reduction in smoke output. Yet there is an alternative history, where the introduction of new technologies allowing for natural gas extraction from the North Sea had a larger influence on this reduction than "mere" legislation.

Did the Clean Air Act spur the development of cleaner energy technologies, or was it a fortuitous coincidence? One need not pick and choose. The Anthropocene is, after all, an *and*thropocene, where the full spectrum of actors can never be fully included on the stage, a Greek chorus of woes, uncertainties, and cascading repercussions. Similarly, the critical neologism-fest over what to call the new epoch need not be seen as a witty one-upping nor a more enlightened worldview but merely an endless algorithmic addition: capitalism *and* patriarchy *and* industry *and* colonialism *and* e-waste *and* extractivism *and* climate change *and* species extinction . . . the list goes on. An era of accumulation provides a whirlpool of factors that overwhelm and exhaust even the most polymath. Successful environmental legislation and campaigns tend to focus on a specific issue: *this* dam project, *this* one chemical, *this* endangered species, *this* rainforest.

Seeing may be believing, but seeing alone is not action toward a more equitable planet. Let the second maxim of the Mediacene join quickly: *representation ≠ reparation.* Finis Dunaway convincingly presents the importance of iconic environmental images for galvanizing activist efforts. From the "Crying Indian" to photos of oil-soaked birds to the recycling logo, images can become popular flash points that crystallize a sense of urgency and outrage. Yet Dunaway also notes how media representation has often promoted surface-level changes that emphasize individual actions and "end of the pipe" remedies that seek not to reduce pollution but relocate it to poor, minority neighborhoods and developing countries. In the case of images that attempt to present global environmental issues, the disconnect only widens between "dramatic spectacles of danger and defilement" and a lack of "broader structural ideas about how to reduce toxicity, diminish greenhouse gas emissions, and envision alternatives to industrial agriculture."[25] As such, the global solidarity conjured through these images is often a shallow one that promotes therapeutic relief for the individual while largely allowing neoliberal capitalism to operate as usual.

If mainstream, mass-media imagery has not proven up to the task, Demos and others seek a form of high-brow, avant-garde, artistic, literary, or other aesthetic *representation* that could politicize the Anthropocene in a way that holds corporate capitalism accountable for its regime of exploitation. For Nixon, writer-activists can employ imaginative writing and other narrative techniques to represent environmental issues facing communities that are normally of little interest to the corporate media. Yet Amitav Ghosh questions whether literary form is up to the task. Ghosh notes the difficulty of promoting ecological understanding that would take account of both gradual, accumulative change and spectacular, cataclysmic events of short duration. Realist fiction especially fails at this task because "the very gestures with which it conjures up reality are actually a concealment of the real."[26] He connects the shortcomings of form and imagination to political failures that have led to global crisis.

While the novel, with its bias toward bourgeois regularity, may be particularly susceptible to these pitfalls, Ghosh's thesis can arguably extend to other aesthetic pursuits. All forms of creativity are invested to some extent in ways of obscuring the means of the medium in favor of greater immersion and engagement. Calling attention to these techniques, however, can lead to innovations. Take impressionism: visible brushstrokes overturned the dominant school of realism that sought to hide the painter's hand. At the same time, impressionist paintings do not reveal the technological changes that facilitated the style: the development of brighter, synthetic paints and the portability of premixed tin tubes that allowed painters to move from the studios to open air. The vibratory aesthetic did not unmask painting but instead created new potentials for the medium that would go on to develop through channels of figuration and abstraction.

Even if the perfect image, film, novel, hypertext, VR experience, or whatever it might be did in fact exist, it would be met with a larger issue. Representation alone is not enough to repair the earth. Media theorists call for ecologically informed representations that can politicize, raise awareness of, and concentrate emotion on environmental issues. But a greater problem may be an *oversaturation* of green messaging within the scientific community, the arts, the news, the film industry—indeed, all industries and walks of life. If observation fulfills an analytic need, representation can fulfill an emotional need to provide a release for climate anxiety and feelings of guilt, shame, or complicity. This therapeutic end can in turn normalize planetary crisis as a persistent part of modern life. Is amplifying environmental representation in the ecology of attention enough?

Media in the twenty-first century has changed drastically from the familiar forms of the twentieth century. If older media forms are found wanting, there is enthusiasm for new media projects that provide immersive experiences of nature and simulations of environmental threats.

The "new media" revolution promises interaction, connectivity, and user-generated experiences. Daniel Reynolds explores the potentials for moving media theory away from the "bounded figures of the body and the medium."[27] Further, this shift refracts backwards to consider how participatory registers exist in older media. Yet before overhastily celebrating new media as a panacea for the alienation and passivity inherent in previous versions of the spectacle, one should consider the limitations to this interaction as well. The third maxim stands as follows: *participation ≠ praxis.*

How much of online activity is truly agentive? While social media has real effects and consequences, much digital interaction provides the illusion of activity from a set of predetermined choices. At the same time, many activist causes have benefited from hashtag organizing, viral videos, and the twitterverse. Maëlle Jacqmarcq argues that digital platforms allow environmental groups to employ "tactical media" strategies, but they also encourage "slacktivism" and "echo chambers" that reinforce the status quo.[28] Further, many online environmental movements inadvertently lead to new forms of consumption and commodification in the name of protecting the planet.[29]

As technology becomes more complex, art can move beyond representing nature to refashioning it. Elizabeth Swanstrom argues for the potentials of the digital to eschew mimetic representation and become "full-frontal aesthetic encounters that demand our active and conscious participation," through which people might in turn realize their "embodied materiality" and "kinship with other living and nonliving things."[30] From this perspective, the way into deeper connection with the nonhuman world is not by rejecting the technological but precisely the opposite. Other media and art forms have not gone far enough: total immersion and active interfacing provide not just a site of contemplation between work and viewer but a lived, participatory experience.

While this premise is sound, the promise that one will reemerge from such an experience with a clearer understanding of positionality and agency seems far from guaranteed. First, the question, why reemerge at all? If technology can provide natural experience but better, why go to the trouble of engaging with real environments, which are notoriously full of inconveniences and discomforts such as travel time, unexpected encounters with bugs or less desirable organisms (other humans included), and inclement weather? Second, the embrace of the metaverse as a greener world by prominent groups such as UNESCO and the World Economic Forum and corporate initiatives like Microsoft's "Planetary Computer," Google's "Earth Engine," and IBM's "Environmental Intelligence Suite," should give pause. Such digital optimism holds that participating is having one's cake and eating it too, but might it not also be the cake which is doing the eating in these scenarios?

It is impossible to generalize a singular experience in the cyber world. What one person finds draining, inane, or a waste of time, another might find meaningful, empowering, and life affirming. Works from scholars such as Sherry Turkle and Stéphane Vial take an important step back to consider the phenomenological repercussions of mediated life. Distinctions between the real and the virtual are less important than the process of perceptual and behavioral habituation that occurs as more users participate in a new technology. For Vial, "the technologies that we live with at a given time condition the very modalities of the possible manifestation of the world at that time."[31] The challenge is to engage with technologies that shape the world while holding them accountable as well.

If the quality of human participation in the digital world varies and exciting new opportunities for art and meaning making clearly exist, it is also worth considering how the nonhuman world participates in these systems. While humans enjoy more enriching experiences and benefit from the virtual's ability to forge new realities, the flora, fauna, and matter that make up ecosystems cannot follow their human counterparts. The nonhuman participates only as resource and referent for immersive media environments.

Rebel Geologies

The Mediacene suggests a global spectacle where the roles of director, producer, actors, and audience are ambiguous at best. The trajectory of media theory from the 1960s on has proven very good at theorizing environment, subjectivity, networks, and the effects of technology on social interaction. Ecological metaphors and analogies abound in contemporary works, but the development of the Anthropocene discourse has arguably thrown a wrinkle into theories of media and mediation. How to define the Anthropocene? It forms part of a chain of nebulous crisis signifiers: from global warming to climate change and now the human epoch. Global warming proved to be too specific. Climate change is perhaps too amorphous; any slight aberration may now be labeled as such. The Anthropocene is a leap beyond these other narratives, inviting consideration of not only climate systems but also the entire scope of environmental afflictions on planet earth. At the same time, the focus on geology grounds the atmospheric narrative both literally and figuratively.

Theory that delighted in postmodernist abstraction and phenomenological wanderings must now contend with the material influence of the *technosphere.* A number of recent works from Jussi Parikka, Sean Cubitt, Richard Maxwell, and Toby Miller have cataloged with agonizing precision a litany of factors such as heavy metals, carbon emissions, rare minerals, toxic manufacturing, planned obsolescence, and exploitative labor conditions that contribute to the material consequences of media and information technology.[32] The result of this focus for Parikka is an alternative media theory that "may not include McLuhan, Kittler, and their like."[33] Behind the push to get rid of these theorists is perhaps an attempt to align media analysis with the scientific rigor of measuring physical reality.

Geologists, meteorologists, and biologists can make clear the reality of human influence on planetary systems, but this tells only half of the story. The other half lies in unpacking the historical, economic, social, and cultural factors that have led to a human epoch of devastating consequences. To this end, the matter of media remains tied to theories of immersion, attention, escapism, affect, virtuality, and imagination from writers such as Guy Debord, Jean Baudrillard, Katherine Hayles, Eugene Thacker, Jonathan Crary, and Donna Haraway, among many others. It is not within the scope or ambition of this book to perfect a unified theory of media, but a comparative approach can help to create a more informed set of tactics for resisting the normalization of planetary demise. The Anthropocene requires a *rebel geology,* drawn from the long accumulation of discourse, philosophy, art, literature, and other forms of human creativity.

A comparative media approach is necessary to get around the monolithic presence of humans as a world-altering force. At its best, this methodology operates under the hope that, in the words of Hayles, "techniques, knowledges, and theorics developed within print traditions can synergistically combine with digital productions to produce and catalyze new kinds of knowledge."[34] A comparative media approach draws knowledge and practice from literary analysis to bear on film, television, video games, electronic literature, and any number of hybrid and polymorphous texts in between. One need not privilege the literary in this approach: newer media forms can equally provide critical techniques and practices to recontextualize older media.

Beyond comparing different types of media, works that seek to comment and reflect on a different medium are of particular interest in this study. Literature can depict psychological aspects of a life lived with television, just as a film can employ techniques of montage or time-lapse to portray the sensation of digital addiction. Works that are self-reflective about the creative process itself provide important lessons for the other things people create, notably waste and pollution. Works of *poioumena*, metafiction, autofiction, the play within the play, self-portraits, films about filmmaking as well as works that portray one form of media in another show the *autopoietic* and *remediated* nature of creativity: a constant and referential system of production and reproduction.[35]

Technology advances in a similar fashion, combining aspects of different mediums to offer more complex systems. Video artist Nam June Paik, for example, developed with engineer Shuya Abe a "Versatile Color TV Synthesizer" that allowed him to play the television like a piano.[36] Comparison becomes harder in this move toward synthesis, what Alenda Chang terms the "transmedial." Focusing on video games, Chang notes not only the blending of film, music, and literature, but also a further articulation of transmedia that explores "the correlation between the real and the virtual, lived life and played life."[37] In this sense, people turn to media not to escape the real world but to better understand their positionality in a larger social milieu.

Comparative and interdisciplinary approaches have long defined work in the environmental humanities, which seeks to build solidarity between distinct disciplines through a shared concern for the earth. Reaching across the aisle to engage with the sciences is crucial for ecocritical works that seek a rigorous understanding of the discourse surrounding crisis. Theories of networks and shared materiality inform articulations of environmental politics and aesthetics from Bruno Latour, Timothy Morton, Jane Bennett, Cary Wolfe, and others who pursue frameworks beyond anthropocentrism such as actor network theory, new materialism, posthumanism, and object-oriented ontology. However, a too associative and eclectic comparison risks becoming an amorphous and arbitrary endeavor.

As persuasive as a comparative approach is for showing continuity and the immersive mediation between humans and the nonhuman world, this book also contends that it is important to resist a too facile ecological aesthetic of holistic connection. The reason for this is that similar logics of interconnection and continuity also inform imperialist and neoliberal narratives of globalization that seek to manage populations and monetize the planet. Attempts to represent these networks can both reveal and obscure global understanding. Media projects operating on a planetary scale participate in the process of *mondialisation* that feeds the hubris of aspiring to understand, control, and manage Earth in its entirety.

To temper a purely free-flowing analysis, this book maintains an engagement with history, temporality, and narrative. The teleological discourses of modernity do not simply flow in one direction. The forward movements of technological progress, political liberation, and planetary knowledge run up against the ebb tide of accumulation, pollution, toxicity, depletion, and scarcity. Environmental justice movements are a constant reminder that progress and protection for some often lead to setbacks and suffering for others. Paying attention to the comparative and the transmedial should illuminate not just connections but also the disconnections, limitations, and boundaries of media.

Through Lines and Dead Ends

The following chapters tend to embrace distraction and multiplicity rather than rehashing the common lament that the digital age is simply an attention deficient mess. DeLillo's diagnosis of "brain fade" and Nixon's bemoaning of the internet's "info-whelm" are ultimately reductive characterizations of the contemporary world. Further, such dismissals indulge in a pastoral nostalgia for a simpler time when attentive practices were deep and sustained, one that scholars in the digital humanities have rightly challenged. Marion Thain, Rita Raley, Malcolm McCullogh, and others seek to work with distraction to call attention to the interfaces and devices that facilitate the wide variety of "readings" available in different forms of media. Thain asks "how we might harness these delicious distractions as attentional devices—and in the process develop much greater self-awareness of the role these technological tools play in our lives, and, crucially, the role we want them to play in our lives."[38]

The academic monograph is a tool no less in need of reevaluation. Writing a book requires a series of mediations that ultimately lead to a fixed text. For better or worse, these paths have been drawn, and authors can only leave it to their readers to construct a multiplicity of forking paths from the one garden. While the subsequent discussion hews somewhat to a historical timeline of technological development (geological time is under consideration, after all), each chapter considers an eclectic set of texts, images, and objects that cater to the attention-deficient reader. Much has been left unexplored and undeveloped as a result. In the tension between connection and disconnection, through lines and dead ends, the reader will hopefully also question the role that writing plays in relation to environmental causes.

Part 1, "Golden Spikes and Splices," lays some groundwork through a longer look at aesthetic experiences and historical processes that have led to environmental issues in the twentieth and twenty-first centuries. Chapter 1, "Aesthetics of Antimediation" considers the virtual spaces of contemplation and creativity as precursors for an anthropogenic world. A longer history of mediation combines Marxist ideas of class struggle with contemporary theories of immersion, interconnection, and networked materiality. To resist the mediation of ecological crisis, ideas of "antimediation" from Henry Louis Gates Jr., Eugene Thacker, and Georges Bataille provide a basis for how artists, writers, and activists can disrupt passive consumption and reveal revolutionary channels. Aesthetic forms can create moments of antimediation by subverting the expectations of a specific medium. An increase of self-awareness in the creative arts occurs alongside a rise in global awareness in the twentieth century with the growth of planetary media, modern meteorology, and international networks.

Chapter 2, "Natural Historicities," considers how imperial expansion around the world set in motion the collapse of history, nature, and aesthetics. An initial discussion of fin-de-siècle aestheticism and the natural sciences surveys artistic responses to an age defined by facts and objectivity. While artists can critique forms of naturalism, governments and institutions rely on statistics and propaganda to naturalize historical destiny. Founded in 1888 to study the products and economic trade of the British Empire, the Imperial Institute's move toward an "Empire Story-Land" in the 1930s with films, theaters, exhibits, and publications sought to create emotional ties between citizens of empire and natural products throughout the world. These state-sanctioned stories of imperialism come under critique in film and fiction such as Christian Nyampeta's *Sometimes It Was Beautiful*, Chimamanda Ngozi Adichie's "The Headstrong Historian," and César Aira's *An Episode in the Life of a Landscape Painter.*

The legacies of imperial media networks continue in the growth of global media conglomerates during post-World War II decolonization. Part 2, "Whole Earth Resistance," looks at the coemergence of media theory and environmental movements in the 1960s and 1970s. Chapter 3, "From Global Village to Anthropocene," explores how holistic visions of a noosphere collide with civil unrest and energy crisis. Marshall McLuhan identifies artworks as "anti-environments" that recontextualize new forms of human sensation. Novels such as Walker Percy's *The Moviegoer* and Jerzy Kosinski's *Being There* present characters seeking to escape the numbness of media culture. From the dirt and stone of Michael Heizer's "Dragged Mass" to the electromagnetism of Nam June Paik's "Global Groove," artists expand the environmental potentials of aesthetic expression. An alternative to mainstream mediation comes from the *Whole Earth Catalog*, a *Walden* for the technological set outlining sustainable consumer culture for the denizens of Silicon Valley and beyond. While alternative lifestyles offer a more sustainable world for a select few, questions remain over how to expand such choices to a global population.

Across the Atlantic, Guy Debord and the Situationist International sought revolutionary applications of emerging media technologies to push back against the separation and consumption promoted by mainstream capitalist spectacle. Chapter 4, "From Ecological *Détournement* to Rebel Geology," explores key terms such as "psychogeography" to understand the influence of surroundings and "détournement" to consider how art can hijack and reroute cultural objects. Patrick Modiano's novel *In the Café of Lost Youth* presents a fictionalized account of the Situationist movement to explore the social repercussions of its techniques. The legacy of the SI after the protests of 1968 continues in the design group Utopie, which attempts to synthesize the arts, architecture, and theory to revolutionize urban experience. In becoming more environmental, creative works face the paradox of turning into immersive technological installations in their own right.

Lessons from the arts and activism of the 1960s and 1970s find new resonance for thinking about environmentalism in the hypernetworked world of the twenty-first century. Part 3, "The Golden Age of (Dis)content," examines digital texts, video games, augmented and virtual reality projects, and artificial intelligence. Chapter 5, "Cyber Saturnalia," draws on contemporary theorists such as Bernard Stiegler and Gilbert Simondon to consider how corporations pacify populations to ensure ongoing consumption and unsustainable economic growth. Unlike the Roman Saturnalia, which occurred once a year, the ecology of media presents constant access to role reversals, binge-watching, and orgiastic indulgence. To recover the Sterculian side of Saturn as a force of regeneration, the chapter considers the joys and consequences of digital life portrayed in Ted Chiang's *Exhalation,* Patricia Lockwood's *No One Is Talking About This,* and Jenny Offill's *Weather.* Beyond incorporating online discourse in traditional print, a growing body of electronic literature composts old techniques to yield fresh forms of interactive texts. A new cry for saturnalian rest arises in the "digital detox" that attempts to rediscover natural authenticity.

Unlike the inverted world of the Saturnalia, chapter 6, "Digital Twins," looks at aspirations to replicate the physical world beyond mere simulation. E. M. Forster's "The Machine Stops" provides a speculative view of life dictated by isolation and automation. Cloud computing presents the fantasy of preservation unmoored from physical infrastructure. Virtual and augmented reality experiences such as Sarah Rosalena's "For Submersion" and Nancy Baker Cahill's "Mushroom Cloud" help to increase empathy with the nonhuman world. Artificial intelligence can make aesthetic interventions of its own and solve logistical questions for mitigating environmental issues. The question remains: Will humanity's digital twin act as a steward for nonhuman life or magnify the self-interest of its fleshy doppelgänger?

What can break through the seamless mediation enjoyed and aspired to by billions globally? The conclusion, "Reality's Return," considers the potential for environmental crises to act as antimediators that disrupt complacency and renew attention to humanitarian need, species extinctions, climate change, and environmental justice for supporting life on a planet in peril. Environmental media in the twenty-first century faces the challenge of recuperating the energy of the real to help people spend less time in spectacle earth and more time engaging with ecological crisis.

PART I

GOLDEN SPIKES AND SPLICES

1

Aesthetics of Antimediation

> Nothing is beautiful, except man alone: all aesthetics rests upon this naïveté, which is its first truth. Let us immediately add the second: nothing is ugly except the degenerating man—and with this the realm of aesthetic judgment is circumscribed.
>
> —Friedrich Nietzsche, *Twilight of the Idols*

If humans have influenced nature and landscape on an epochal scale, how have environmental aesthetics changed as well? What does it mean to make art in an increasingly artificial world? Can one really maintain disinterestedness in the face of complicity? Can environmental aesthetics take on ethical and political imperatives that apportion responsibility for disaster and crisis? While most environmentalists agree that solutions to ecological crises are complicated and require an interdisciplinary effort, not enough attention is given to the processes of mediation necessary to navigate this global collaboration.

Lorraine Daston's work on how ideas of what is natural inform society's moral and political codes offers one potential starting point: "Only when these natural disorders shade into human culpability does outrage tinge the passions of the unnatural. If the monster is believed to be the issue of a sinful union, if the drought is believed to be due to greedy land grabbers . . . only then can indignation be unleashed."[1] Cases of human accountability have largely galvanized environmental movements. From the 1969 Santa Barbara Oil Spill to 2010's Deepwater Horizon fiasco, corporate negligence has crystalized anti-industrial sentiment and mobilized the masses. Environmental justice projects and NIMBY organizers find success by identifying toxicity in neighborhoods and mitigating these situations through class-action lawsuits and court-mandated clean-up efforts. This is mediation at work, moving from the perception of the human destruction of ecosystems to direct efforts to restore self and surroundings.

Aesthetic works also do well to heed a specificity of focus if they wish to have an impact. Roger Payne's *Songs of the Humpback Whale,* for example, inspired movements to protect ocean life and launched a wave of "earth music" led by Paul Winter and others who sought to intertwine the voices of humans and animals in compelling arrangements. For his 1980 album *Callings,* Winter describes an encounter with a sea-lion pup in Baja California that would not only rouse changes to his saxophone playing but also rewire his ontological state of being. Describing the work, he writes, "The seal, for me, is a mediator: between the sea creatures and humans, between the land and the sea. I feel a kinship. I want this music to reawaken and celebrate this kinship."[2] Here mediation is an expansive process, moving from a direct encounter to a larger awareness of connection. The hope is that the resulting aesthetic work can replicate this process in the listener who will in turn reach out to political campaigns to stop seal hunting and change consumption habits to limit the demand for sealskin.

In the twenty-first century, direct links between perceiving victims and external culprits still exist, but an increasingly murky entanglement surrounds the less sensational and more dispersed risks of planetary change. The Anthropocene is not merely a spill that can be cleaned up, atoned for, or balanced through a monetary payout. In the place of a passionate outcry that can be pacified through direct action, there exists a low-level hum of anxiety, suffering, guilt, shame, and denial that cannot be resolved precisely because it is difficult to express in the first place. Faced with these feelings, humans take refuge in the virtual escapism available through media technologies or the green cocoon of organic products, self-sufficient dwelling, and superfood supplements.

One result of thinking about human impacts at a species level is to extend some amount of culpability to everyone. Some commenters see this as a good thing. Others see this deflationary gesture as a smokescreen to cover for the real culprits: the military-industrial complex and its corporate handmaidens who turn to greenwashing to make superficial gestures toward sustainability. Would it be possible to hold the worst offenders accountable while acknowledging a species-wide complicity? Even the most environmentally conscious leave their footprints. Further, might a sense of universal accountability help to increase the widespread democratic participation needed to enact legislative, economic, and political change? To "stay with the trouble," in Donna Haraway's formulation, humans need new forms of aesthetic mediation that can drive changes to species-being on a global scale.

The Sage in the Cave

Efforts to confront the new scale of ecological changes are already underway in many sectors of the arts, academia, and media. Berlin's Haus der Kulturen der Welt's Anthropocene Project, for example, calls for "a new sense of amazement at the wonder of the Earth."[3] The artists and writers involved in this project take it as axiomatic that this new aesthetic sensibility should be firmly rooted in the material. Indeed, many have embraced the arrival of the Anthropocene for its imperative to give up postmodern abstraction and return to reality. While it may be true that "the materiality of the Anthropocene turns out to be its most powerful feature," it remains to be seen how this power exerts itself over human activity.[4]

In the new materialist world, nothing is solid or separate. Everything bleeds into everything; order and hierarchy give way to constant flux and dissolution. After generations spent constructing and modernizing civilization to keep nature out, the nonhuman returns to revel in an analog continuum of transmissions and affective oozing. Humans, in a tremulous spasm, give up their enlightened position and recognize their true place among the gunk and slime of the physical world: a creatural writhing with carbon-based comrades.

This focus on what is literally in one's hand—the metals, chemicals, and polymers that make up devices—drives the media geology of Jussi Parikka's "anthrobscene," which brings attention to the "a-signifying, semiotic matter" of rare earth minerals and futuristic alloys, extending in turn to brutal mining corporations and the exploited labor of neoliberal manufacturing plants.[5] When looking at one of the over seven billion smart phones currently in use worldwide, the user should see beyond the glossy perfection of pixelated images to notice all of its physical flaws: the cracked screen, the smudges on the glass surface, the grated aluminum corners. The mind will, in turn, travel all around the world to the mines of Tanzania, the lithium farms in Chile, the manufacturing plants in China.

If this mental voyage is not enough, one should also feel a profound depletion day after day trying to constantly recharge a dwindling battery that expires faster and faster over time. It is like Sylvia Plath's "Lady Lazarus," who reminds the reader that

> There is a charge
> For the eyeing of my scars, there is a charge
> For the hearing of my heart—
> It really goes.
>
> And there is a charge, a very large charge
> For a word or a touch
> Or a bit of blood
>
> Or a piece of my hair or my clothes.[6]

The sky-blue glow of the LED screen. The silent warmth of hyperefficient processing chips. The zero-emission vehicle. There is a charge, too, for these seemingly magical objects.

Despite this focus on matter, the clarity of "reality" in contemporary life is far from being a fait accompli. Even if it were, simply pointing to the real is unlikely to change human behavior. Most people are aware of the disheveled state of their environment. The threats of air pollution, trash, chemicals, and viruses are a constant reminder of reality's razor-sharp edge. The insistent *presence* of material reality drives escapism and flight into the virtual realm, the mediated world of online existence, viewable content, and produced poetics.

It is helpful, however, to think through how concepts and constructions of reality have changed over time. Many material analyses blame the death-binding machinations of industrial capitalism for constricting *life* itself. Andreas Weber in his work on "enlivenment" outlines how petromodernity colonizes reality. He calls for reinvigorating life through a return to the commons. Against a biomechanical order that reduces living matter to objects and resources for economic exploitation, Weber proposes an epigenetic model of biological activity where all organisms are agentive beings contributing to world-making collaborative exchange.[7]

While Weber's articulation of enlivenment is crucial for stressing the agentive potentials of *reality,* his conclusions seem far from guaranteed. Many materialists reach this impasse: in their view, a new ecological awareness and way of life will emerge if everyone simply recognizes the interconnection of matter. Timothy Morton, for example, proposes a process of "rocking" back and forth with the nonhuman to create the good vibrations that will heal the fractured world.[8] People need only see themselves in the grass, the tarantulas, and the sea anemones to transform their surroundings. A strange dogma creeps back into the liberatory explication of liveliness: individuals are agentive, their decisions matter at every moment, and yet, for these reasons, everyone should think the same way about how to coexist with matter on Earth. A more robust articulation of ecological ethics must consider the wide range of cultural attitudes toward the nonhuman world.

Also absent from most of these material analyses is the recognition that many people *prefer* to engage in a world of screens and virtual mediations. Weber champions a return to the commons but fails to acknowledge that the contemporary commons is becoming a digital one, where humans meet through the mediated layers of social algorithms, anonymous personas, and stratified speaker channels. Of course, these domains have their material infrastructures as well, but the communion with digital beings should be distinguished from their fleshy counterparts. It does not make sense to consider the interaction with an avatar merely an encounter with plastics, liquid crystals, and low-wattage energy from coal-fired power plants. Something else is at play: an abstraction, an imaginary form, a projection of self.

What would it entail to rethink the flourishing of materialist critique through these mediated encounters? Katherine Hayles also turns to a discussion of epigenetic evolution to understand the technogenesis occurring as a feedback loop between people and technology. As the speed of life increases, the technophilic become motivated to "modify their environments so that they become yet more information intensive."[9] While Hayles celebrates the new potentials of knowledge and interaction through technological changes to the environment, the Anthropocene is a testament to the other reality of technogenesis: the accelerating feedback loop of disorder, toxicity, and precarity.

Material bias is built into discussions of the Anthropocene for the simple reason that this is how geologists measure its existence: CO_2 levels, ocean PH, soil phosphorous, heavy metals, radiation, chemicals, and so on. Histories and narratives reflect this bias, pointing to technological shifts such as the nuclear bomb, the Haber-Bosch process, or the steam engine. But beyond these technological landmarks lies a deeper existential discussion of human activity, the way in which people pursue separation from the environment, other species, and, finally, each other.

Thought, consciousness, and theory hold the seeds of environmental neglect. Virtuality, boredom, detachment, and apathy have histories much longer than immersive digital technologies. Abstraction from the physical world is as old as imagination itself and inheres in older technologies such as books and paintings. The aesthetic, spiritual, and material dynamics at play in simple, mediated abstraction can be seen, for example, in a work such as Lucas Cranach the Elder's 1526 painting *Cardinal Albrecht of Brandenburg as Saint Jerome* (see fig. 1).

Second only to St. Augustine for his output in Latin Christian writing, Jerome's industriousness was a lifelong pursuit that led him to asceticism, a scholarly retreat in Bethlehem to the cave where Jesus was purportedly born, and a serious vitamin A deficiency. His turn toward religious themes came not with any grand illumination, but rather from his experience visiting the darkness of the crypt. In this pure, silent void, with its affirmation of mortality, Jerome began a creative process as a deliberate actualization of his presence in the world. Although he didn't know it at the time, his translation of the Bible would go on to be the most influential text in the Western world for the next thousand years. As the perfect passive participle of the Latin *vulgo,* to broadcast and make known, the Vulgate Bible lived up to its name.[10] Vulgate, of course, also connotes *vulgatus* and the vulgar, to prostitute or make common.

Cranach the Elder was a friend and sympathizer of Martin Luther. He painted portraits of the reformer and printed Luther's German translation of the Bible with custom woodcut illustrations. Martin Luther was not impressed by Jerome's asceticism and did not see his Vulgate Bible as authoritative. Luther took it upon himself to create a new translation in the common tongue. Cardinal Albrecht chose this portrait as a direct rebuke to Luther's newly translated Bible.[11] Although Albrecht started his career in a liberal vein and took interest in the Protestantism of Martin Luther, he returned firmly to Catholicism after the failed Peasant's War of 1525, which led to the brutal slaughter of one hundred thousand Germans. Preoccupied with his place in the afterlife, Albrecht secured thousands of relics and dozens of sacred skeletons in his reliquary, Hallesches Heiltum. Much of these material goods were acquired with funds raised through the sale of indulgences, a practice that spurred reformation leaders to

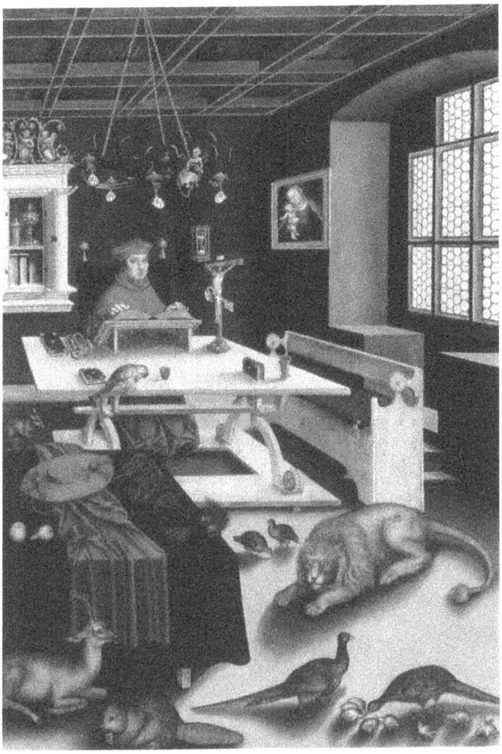

FIG. 1. *Cardinal Albrecht of Brandenburg as Saint Jerome,* Lucas Cranach the Elder, 1526. Oil on wood panel, 45¼ × 35$^{1}/_{16}$ inches. (Collection of the John and Mable Ringling Museum of Art, the State Art Museum of Florida, a division of Florida State University; bequest of John Ringling, 1936, SN308)

break from the Catholic Church. Albrecht pursued the construction of a larger church to hold grander, more spectacular masses that drew overflowing crowds, but ultimately his debts proved too large, and he left the town of Halle that he had done so much to shape.[12]

Cranach, then, was in an interesting position for this commission. His portrait recalls Albrecht Dürer's iconic 1515 engraving *Saint Jerome in His Study.* Instead of deeply absorbed focus, here the figure gazes vacantly into space. Is it boredom? Contemplation? Epiphany? Or mere distraction? Compared to Dürer's presentation of an austere study with snoozing lion and dog, and far from Jerome's modest cave in Bethlehem, the cardinal's study is luxurious and overstuffed with a menagerie of animals, art, adornments, and even an infant angel dangling from the lavish antler chandelier. Might not there be a slight look of guilt or dread in the far-off gaze, a recognition of the futility of these material shores against a changing world order?

While Cardinal Albrecht brokered the purchase of pardons with esoteric aesthetic objects, today a new form of indulgence is taking place in the form of carbon offsets, where corporations mitigate their emissions by paying to plant trees or develop other forms of carbon capture. While such efforts are better than nothing, they bely another materialist attitude toward climate change: there is a way to buy off the problem, to create new markets where others may make their fortunes on the carbon trade exchange. The materialism of geoengineering, "sustainable" development, and "clean" energy markets should give pause and caution to activist and critical projects couched solely in materialist terms. At its best, art can reveal the paradoxes of material mediation. Cranach's painting goes beyond a mere portrait of Albrecht to articulate his culpability within a larger world of spiritual, economic, and political forces.

The painting caught the attention of John Ringling, one of the brothers behind the circus spectacle marketed as the "the Greatest Show on Earth" and currently resides in the Ringling Museum on his sprawling estate in Sarasota, Florida, dangerously close to the rising Gulf of Mexico. What attracted Ringling to this work? Was it a random acquisition among Renaissance masters? As one of the preeminent showmen who brought entertainment to the masses, did Ringling feel a kinship with Jerome, the cave-dwelling saint who translated the word of God? He may also have sympathized with Albrecht, a fellow collector of oddities, animals, relics, and magical objects, whose attempts to sway, educate, and uplift the masses often required the paradox of unsound financial activity, exploitation, and the constant need to find more spectacular forms of communication. Or did he see him as a fellow clown?

Alongside the solemnity of Mass, the circus is instructive for thinking about the compression of planetary life that occurs in the name of spectacular consumption. An ephemeral event, lasting only a day, the Ringling Bros. and Barnum & Bailey circus was a veritable travelling microcosm of not only species diversity but also of humanity's self-organization into a hierarchical species. The spectacle display or "spec" lined up performers and animals alike—dressed in lavish costumes playing on historic and literary figures—to parade into the big tent, giving a taste of the wonders to come.

In its golden age, from 1919 to 1938, traveling from town to town on the railroad tracks, the circus employed over 1,300 workers and performers, setting up the big tent to perform to some 13,000 people, repeated

150 times over in a season. Each day, the cookhouse would go through 2 barrels of sugar, 285 pounds of butter, 226 dozen eggs, 1,300 pounds of vegetables, 2,470 pounds of meat, and 2,200 loaves of bread, among other items. Those higher up in the company and prize performers garnered private tents and rest areas. Over 800 animals traveled with the circus from different continents, a menagerie of hippos, giraffes, polar bears, and kangaroos, among others. Along with these exotics, over 400 baggage horses helped move everything around and lift the tents. Before the final acts of the evening show, workers were already well underway packing everything back up to move on to the next town. Ringling traveled in his own custom-designed private coach, complete with kitchen, bathroom, and bedroom.[13]

The Revenge of Extremes

What to make of a strong uptick in new materialist philosophies that call for *anthrodecentrism* at a time when humans have more influence over the habitats of the world than ever before? A knee-jerk reaction is that this is sheer escapism and strange denial. Humanists were quick to push back against a unified *Anthropos* for the disservice such a concept does to the diversity of cultures and attitudes toward the environment. Some see the Anthropocene as a pivotal moment for activists to move beyond "ethnic environmentalism" by foregrounding matter as a storyteller.[14] Others maintain the necessity to "move colonialism and race from the margin to the center as the primary factors defining the Anthropocene."[15] In regards to the two-headed dragon of environmental and economic crisis, Rob Nixon poses the question succinctly: "How do we tell the story of *Homo sapiens* as an Anthropocene actor in the aggregate, while also insisting that the grand species narrative be disaggregated to reflect the radically divergent impacts that different communities have had on planetary chemistry?"[16] The comparison of geological sedimentary layers and social stratification must go beyond the metaphoric to account for this great divergence of agency.

In a seeming escape from the subjective bind, critics turn to the materiality of bodies themselves, calling attention to the imperialist and capitalist logic that views slave labor and natural resources as one and the same. Jamaica Kincaid puts this starkly in *A Small Place,* when she asks, "Do you know why people like me are shy about being capitalists? Well, it's because we, for as long as we have known you, *were* capital, like bales of cotton and sacks of sugar, and you were the commanding, cruel capitalists."[17] Kincaid's use of second-person narration throughout her jeremiad reinforces the distance between "you" and "I," "us" and "them," all members of the same species. Kathryn Yusoff makes a similar point, that the "move toward a more expansive notion of humanity must be made with care" so as to avoid reinforcing the "violent homogenization of subjective affects and material possibilities" brought about by imperial geology.[18] Teresa Brennan also details a process where laborers are "squeezed into an alignment with nature" by a capitalist order that views both in terms of energy content.[19] As Kincaid points out, the entanglement brought about by colonial conquest, slavery, and capitalist exploitation is an irrevocable one. There is no undoing the rift, no going back to a time of ancient civilizations or a more animalistic existence.

Instead of anthrodecentrism, this book pursues a theory of ecological mediation. This does not mean, however, that a materialist approach goes out the window. Indeed, mediation has been theorized extensively in relation to new materialism by Sean Cubitt, who argues that "mediation names the material processes connecting human and nonhuman events."[20] While Cubitt acknowledges the social elements that lead to inequalities within this mediation, he opts to pursue an analysis of "the materials that media are made of."[21] Recognizing materialist mediation is easy; however, articulating the mediation is difficult. McKenzie Wark elaborates on the factors at play: "The situation which mediates between and enables the cut that delineates object from subject is where we find sensation, apparatus, and the labor point of view."[22] Wark's version of mediation includes both immanent and external factors surrounding any given relationship between the human and the nonhuman. His inclusion of labor points to older theories of materialism that continue to dictate interaction with the natural world.

Mediation in relation to ontology, a sense of being in the world, enjoys a long tradition, reaching back to Aristotle, and no doubt further. Emmanuel Alloa locates a set of related terms (*mesotês, meson,* and *metaxy*) that Aristotle employs to situate the "plurivocity of being" alongside the "problem of mediacy."[23] Aristotle famously expounds on the middle path, the mean, and developing a sense of moderation between two extremes. Alloa connects this ethical project to mediacy on an aesthetic level linked to perception and sensations. Aristotle also distinguishes between immediate senses such as touch and taste and mediate senses that operate *through* another medium. Sight, hearing, and smell become noble senses, as they are removed from bodily need, but they still require the transforming mediums of air and water to connect internal organ to external object.

It is within this system of mediation that Aristotle situates human interaction with the world. *On the Soul* posits that "every living being assimilates its environment in order to sustain itself."[24] This is a process both material and immaterial: eating transforms matter from the environment while knowledge leaves matter intact through an assimilation of form. Herein lies potential and plasticity. Alloa explains the crux of ontological mediation: "Mediacy indicates the capacity to *take on the form* of something *without being (it).*"[25] Rethinking this classical model of media theory in the Anthropocene has profound implications. For one, people no longer assimilate the environment in unidirectional fashion. Environments increasingly assimilate the overwhelming byproducts of human activity. For another, the pure potentiality of medial interaction is less clear: air and water are no longer simple diaphanous mediums, but are shot through with the matter of pollution. John Durham Peters outlines an elemental theory of media as "ensembles of nature and culture" to understand the "enormous pressures that human beings are exerting on sea, earth, sky, and all that dwells in them."[26] Citing the etymology of "aura" as "a gentle breeze or breath," Tobias Menely calls attention to the "auratic quality of air, or the atmospheric quality of aura," in order to diagnose the catastrophe of the present as an "accretion of the past, a thickening of the air."[27] What remains is entanglement: people alter environments, and these altered environments are in turn changing the composition of human bodies.

Mediating this entanglement presents its own challenges. Eugene Thacker explains the resulting bind of navigating the brave new world of planetary crisis: "The world in which we find ourselves today constantly challenges our conventional ideas about media and mediation. In the midst of planetary disasters both human-made and nonhuman oriented, we find we are in the position of recording and documenting events that increasingly slip from our comprehension, events that we can only define using vast phrases like 'global climate change' or 'planetary extinction.' We are living in the very world that we stand apart from in order that we may record, document, and mediate it."[28] This apt description can be taken even further to question whether standing apart is even possible. Aesthetic distance is no longer tenable in an epoch where humans encounter the ramifications of their actions at every turn. One way back into the aesthetic is to disavow disinterestedness and double down on affective connection. Kyle Bladow and Jennifer Ladino explore the emotional registers of life, noting how a flood of environmental media leads to both negative responses such as numbness, fatigue, and grief as well as positive ones such as hope and motivation.[29]

Mediation becomes a moving target as planetary systems shift and take on autopoietic processes of change. The middle path is increasingly difficult to walk in an epoch of extremes. While natural disasters and environmental catastrophes have always influenced civilizations, the mildness of the Holocene created a false sense of security in the natural world. As Richard Marzec points out, "The very passivity of the Holocene has influenced the forms of narrative and storytelling to which we have grown accustomed."[30] The challenge is to reassess the narratives, philosophies, and aesthetics that arose within ten thousand years of relative calm.

The Snug of Mediation

To gain a better understanding of human-dominated ecosystems, new materialist theories of anthrodecentrism can be thought in reverse toward the anthropocentric materialism of Marxist mediation. The result, hopefully, will be a materialist theory of mediation that can apply to human-centered questions of environmental justice as well as more holistic views of ecological interdependence. Mediation occurs on multiple levels: between the large and the small, the West and the rest, the corporate and the individual, the public and the private. Mediation occurs in loops between humans and technology, technology and nature, nature and culture. Mediation might lead to practical efforts, restitution, preservation. Or it might operate at emotional and empathetic levels. Mediation is needed for the mediators. *Quis custodiet ipsos custodes?* Who watches the watchmen?

Mediators are fundamental to what Gilles Deleuze describes as the "relations of mutual resonance and exchange" taking place between philosophy, art, and science: "Creation's all about mediators. Without them nothing happens. They can be people—for a philosopher, artists or scientists; for a scientist, philosophers or artists—but things too, even plants or animals."[31] Mediators should be specific but also open to indirect and free associations. "The concept," says Karl Marx, "is that which mediates between form and content."[32] For Donna Haraway, mediation must be constitutive in a way that "profoundly interrogates the received structure of that concept."[33] To what extent can theories of mediation with the physical world be thought through mediation with media itself: literature, art, film, video games, the internet? Raymond Williams invokes Theodor Adorno to argue that artworks do not gain mediation through the presence of a viewer; rather, art is itself already mediated through its production.[34]

Media in its contemporary sense of the collective of workers and infrastructure comprising mass communication is a relatively new consideration. Extensively introduced in 1923's *Advertising and Selling*, featuring articles from the Associated Advertising Clubs of the World, mass media falls to the purview of economics and global marketplaces.[35] Mediation, on the other hand, enjoys a much longer etymology, stretching back to the Middle English translation of Ranulf Higden's *Polychronicon*, where "mediacioun" by Robert, Duke of Normandy allows for "pease" between Malcolyn, King of Scots and William the Conqueror.[36] Significantly, this is less of a concession between two parties than a capitulation to obedience. Nevertheless, the basics of mediation are here established: two parties avoid escalation through the intercession of a third, supposedly neutral, overseer. The earliest sense of mediation from Old French is the act of "division in two." This dualism arguably inheres to later senses of mediation, even as it approaches its most holistic senses of immersion and material saturation. The result of this ontological dualism is to point out that there is no perfect mediation, no completely neutral third party to decide what is just and what is not.

It is an ongoing and imperfect sense of mediation that informs a Marxist understanding of social relations. In this analysis, humans and nature meet through labor. Marx writes in relation to the commodity that labor is "an eternal natural necessity which mediates the metabolism between man and nature, and therefore human life itself."[37] But as labor creates commodities and surplus values, humans become alienated from nature through the abstractions of capital. Marx develops his philosophy of mediation most extensively in his posthumously published *Critique of Hegel's Philosophy of Right*, which finds fault in Hegel's use of abstract dialectics to naturalize socioeconomic relations between monarchs, legislators, estates, merchants, and agricultural families.

Against Hegelian speculation, Marx wants to insist on *actuality:* circumstance, caprice, personal choice, and economic factors. Marx provides an interesting figure for these considerations in part 5 of the *Critique,* which focuses on Hegel's discussion of the Estates. Marx describes an absurdist triangle between the sovereign, the executive, and the estates where the mean between two extremes inevitably becomes another extreme to one of the three parties. In a comical and convincing passage, Marx writes:

> This is a kind of mutual reconciliation society. It is as if a man stepped between two opponents, only to have one of them immediately step between the mediator and the other opponent. It is like the story of the man and wife who quarreled and the doctor who wished to mediate between them, whereupon the wife soon had to step between the doctor and her husband, and then the husband between his wife and the doctor. It is like the lion in *A Midsummer Night's Dream* who exclaims: "I am the lion, and I am not the lion, but Snug." So here each extreme is sometimes the lion of opposition and sometimes the Snug of mediation.[38]

Marx presents a series of bifurcating mediations somewhat akin to Zeno's paradox or a bizarro Bavarian waltz. Something like this process tends to happen when interventions are made into ecosystems. The elimination of wolves throughout the American West, for example, while done to ostensibly protect wildlife, led to a cascading host of issues in the balance of flora and fauna that continue to be felt to this day.

Marx's reference to Shakespeare presents an opportunity to move from mediation as a political process to mediation as a state of being in relation to art. *A Midsummer Night's Dream* presents the conflict between marriage as a political alliance and love as a harmonious promotion of happiness. The drama hinges on a play-within-the-play, a production of *Pyramus and Thisbe* that a group of Athenian workers perform at a three-pronged wedding ceremony. Snug, a minor character, is a joiner, or a carpenter, which also signifies a process of mediation: connecting one piece of wood to another in a seamless fashion (see fig. 2).

The quote Marx attributes to Snug here is not quite right. Snug himself seems to flub the line a bit, but the essence of this disclaimer is proposed by Bottom in act 3, scene 1, as the tradesmen discuss taking precautions so that the play appears real, but not *too* real. They discuss how to simulate

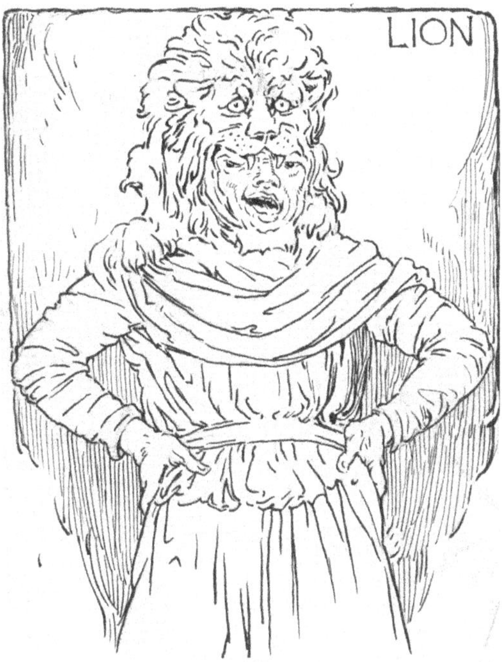

FIG. 2. *Lion,* Louis Rhead, from Charles and Mary Lamb, *Tales from Shakespeare,* 1807. (Folger Shakespeare Library, ART Box R469 no. 65)

moonlight, among other stage tricks, and want to find a balance of titillation without actual fear or shock. Marx may have been drawn to this merry band because they represent working-class Athenians. Theseus, for his part, finds entertainment in this play because it is not too precious: the crude production makes it more entertaining and perhaps, paradoxically, true.

Whatever the case, Snug's own identification and distinction with the role and with the lion itself provides an interesting model for thinking through mediation and materialist naturalism. One should be able to say, "I am the lion, but I am also distinct from the lion. I am connected to my surroundings through a continuous materiality, but I have also constructed an identity and understanding of myself as an individual." There is no need to choose between these positions, and a larger ethical approach to life may depend on a collectivity of Snugs. Commenting on the ability to act ethically toward living systems without assimilating them, Robin Attfield writes, "It is our own very distinctness that gives our ethical convictions their role" and promotes "the kind of ethics which can weigh up diverse and sometimes conflicting concerns and priorities, such as present interests and future ones, or those of different species."[39] To snug is to lie close, to nestle, to rub up against another. This is done not to merge, not to emphasize a fundamental unity or holism, but quite simply to stay warm, to survive.

Antimediators

Marx's snug of mediation turns out to be closer to a form of *antimediation:* a breakdown in communication that reveals the fundamental impossibility of resolution. Such an impasse provides an important model for resisting the fluid connection to media, devices, and attention economies. Environmental activism often begins with just such a feeling of disruption to the natural order. What was once a neutral or benevolent space becomes changed, charged with a threat. Such moments may be more subtle as well: Rachel Carson's *Silent Spring* found its impact by calling attention to the disappearance of birdsongs long taken for granted in American agropastoral communities. Similarly, environmental toxins and the slow violence of climate change may be hard to sense at first.

Theoretical models for antimediation go by many names: the anti-environment of Marshall McLuhan, the *détournement* of Guy Debord, the *via negativa* of Georges Bataille. Henry Louis Gates Jr. identifies antimediation in "monkey tales" as a form of play between Monkey, Lion, and Elephant. "Monkey," Gates explains, "is a term of (anti)mediation, as are all trickster figures, between two forces he seeks to oppose for his own contentious purposes, and then to reconcile."[40] Antimediation may be mediation in disguise. The antimediation of the Monkey proceeds through the inventive and mock-abusive rhetorical strategies of vernacular. Language, as Gates points out, is itself the great mediator, allowing people to understand one another, to advance civilization, and resolve disputes. If language becomes hegemonic, however, cultural development may languish as a result. Gates identifies the process of "signifyin'" to reintroduce linguistic variation and challenge dominant power structures.

Translators and interpreters inhabit the roles of linguistic mediators that allow cultures and nations to interact with one another beyond language barriers, but there is no such thing as a perfect translation. While humans increasingly rely on technology to communicate, Eugene Thacker and Yves Citton invoke antimediation to describe the ways in which media may be used to connect to the spiritual world of ghosts, paranormal activity, alien signals, and other spooky phenomena. This antimediation underlines the simple but fundamental notion that "*media are not human*."[41] As such, something is inevitably lost in any process of exchange. Thacker and Citton invoke Bataille's theory of "negative mediation," which posits a gulf or abyss between sender and receiver that ultimately leads to a breakdown of communication.

Systems designed to connect the world and distribute data become nonhuman, Deleuzian assemblages of antimediation. Alexander Galloway explores the classical figure of the Furies, which he describes as "quite literally anti-media," to analyze the tension between the rise of information networks and the "pure energy of antagonism known as the asymmetrical threat" found in rebellion, guerrilla resistance, and insurgence.[42] Environmental crisis presents an escalation in mediations between humans and the nonhuman. Galloway's infuriation is a compelling vision of the challenge facing the classical model of mediated ontology: "Swarms and systems threaten the sanctity of the human more than animals or things or ghosts. They violently reduce mind to matter, disseminating consciousness and causality into a frenzy of discrete, autonomous agents, each with their own micro functions."[43] This dissolution is the opposite of Althusserian interpellation. Instead of being called into a society by internalizing a value system, antimediation is a violent feeling of disavowal, noninclusion, or being out of place. It sticks out like a sore thumb.

An aporia that presents no exit, a détournement that questions not just a perceived object but the sense of perception itself, a feeling of nausea, withdrawal, or unease. If mediation is a dialectical process, antimediation invokes Walter Benjamin's dialectics at a standstill. Whereas mediation may be a long, drawn-out process, antimediation is a state difficult and dangerous to inhabit. Once felt, a person becomes eager to remediate, to plug back in, to return to the stream of life. Too long a period of disconnection leads to chaos, depression, malaise. For this reason, moments of reprieve are often built into civil society. Beyond the circus lies the Saturnalia, a deliberate time of role reversal, indulgence, and breaking of taboos that is sanctioned momentarily to ultimately reinforce the status quo.

Moments of antimediation can be found in books, music, movies, artworks, or on the internet. The Irish playwright J. M. Synge details a defining moment that occurred to him while reading Charles Darwin's *On the Origin of Species:* "It opened in my hands at a passage where he asks how can we explain the similarity between a man's hand and a bird's or bat's wings except by evolution. I flung the book aside and rushed out into the open air . . . the sky seemed to have lost its blue and the grass its green."[44] In this passage, Synge describes being unprepared to receive knowledge, unready to see the world as well as his own body in a different light. Nature does not provide a vibrant affirmation of his connection with the living world; instead, the surrounding environment becomes static, lifeless. Antimediation may feel this way: a dead phone, a broken link, a missed flight. Sometimes lack of communication says more than words.

Part of Synge's experience is the shock of finding something in Darwin's writing that he was not expecting. Extending concepts such as defamiliarization and détournement to consider how mediums themselves may be unsettled and rethought, antimediation plays on the expectations of genre and formal norms. In the film *Chasing Ice,* for example, the unveiling of time-lapse recordings of melting glaciers ruptures the viewer's experience. The effect is heightened in relation to the first hour of the film, which chronicles James Balog and his team's efforts to place cameras at glacier sites around the world to capture the growth and thaw of ice over a three-year period. The film proceeds with all of the hallmarks that viewers have come to expect from documentary filmmaking: background information, interviews, setbacks, troubleshooting, personal struggles. The time-lapse, while still very much a product of film and media technology, completely subverts this style to capture the scale and intensity of glacial change. What Deleuze terms the "auto-temporality" of images effectively brings the ice to life in a strange dance of froth and decay to reveal something akin to suffering.[45] The viewer, in turn, jolts out of the complacent consumption of a nature documentary into something more spectral and haunting: a sublime understanding of climate change in relation to time and space.

Moments of antimediation are becoming harder to achieve as media in the twenty-first century increasingly avoids the trappings of ornament in favor of direct address and the transmission of firsthand experience. Anna Kornbluh describes a style of immediacy that takes a stance of "anti-mediation" to get at "real life."[46] This style is both suited to and symptom of an immediacy that people face from the threats of climate change. Such an aesthetics is, of course, also mediation, and a form that proves particularly paralyzing in the face of a late capitalist global system that threatens planetary well-being.

In this world of immediacy, the ultimate antimediator may be the return of nature itself, or, rather, an anthropogenic nature made more threatening by human activity. Natural disasters, extinctions, and ecosystem collapse create voids in the continuously mediated ecology of life systems. These are events that resist representation. Some critics maligned the 2024 popcorn flick *Twisters* for failing to more overtly reference climate change, but it did provide a fourth wall break by having a tornado rip out the screen of a movie theater to present the bewildered audience with a front row seat to the real spectacle they crave. If a film within a film is a bit heavy handed, smaller, less devastating changes can also provoke a profound sense of disconnection. The artist Nancy Baker Cahill—whose augmented reality installation *Mushroom Cloud* is discussed in greater detail at the end of this book—describes how seeing an East Coast sky turn orange due to smoke from forest fires all the way on the West Coast created a rift in her senses and a reevaluation of her artwork in relation to place and planet.[47]

Changes to climate and seasonal patterns register in anecdotal experience, leading to a growing sense of the uncanny. These days, just as it is becoming harder and harder to ignore climate change, one must work equally hard to avoid technological mediation. It is increasingly difficult to go "off-grid," to find a space devoid of cellular signals and information waves. Antimediators must contend with what Augé terms *mondialisation*, referring to both the extension of global communication networks as well as the increasing growth of planetary awareness and consciousness.[48]

Orchestrating the Globe

Mediation requires a certain frame of detachment and self-referentiality. The play-within-the-play of *A Midsummer Night's Dream* sets up a space from which to remediate the actions of the play, which itself mediates the trials and tribulations of lived experience. An allegorical strain enters: the human epoch becomes a self-reflexive commentary on how accumulation reflects values, actions, and choices. Elizabeth DeLoughery fruitfully explores a modern allegorical mode that engages nature as history to narrativize the Anthropocene.[49] But beyond allegory what this study finds most useful may more accurately be termed *poioumena:* works of art, media, and literature that are self-referential and critical about the process of creation itself. This reflective engagement with the creative process may in turn provide insights on the metacreation of an anthropogenic world. Nor is this a one-to-one correlation. The work created within the work can have the tyrannical and unexpected consequences of the antimediator, something Flann O'Brien masterfully explored in *At Swim-Two-Birds.*

To conclude in the Shakespearean vein, literary critics often cite *The Tempest* as a starting point for thinking about representations of an anthropogenic world. The play is read and performed as a piece of metatheater, where Shakespeare explores the act of world creation made possible by the theatrical arts. The island becomes both a no-man's-land and an every-man's-land as it represents a spectrum from the idealized golden age of Gonzalo to the mystic enchantment of Ariel to the local knowledge of Caliban. It also sets the stage for discussions of land ownership, private property, colonization, and liberation.

If *The Tempest* is arguably one of the more structurally sound works of Shakespeare, the dramatic tension is largely propped up by elaborate staging and visual effects (see fig. 3). Prospero explains as much to Miranda in the second scene:

> 'Tis time
> I should inform thee farther. Lend thy hand
> And pluck my magic garment from me. So,
> Lie there, my art. Wipe thou thine eyes; have comfort.
> The direful spectacle of the wrack, which touched
> The very virtue of compassion in thee,
> I have with such provision in mine art
> So safely ordered that there is no soul—
> No, not so much perdition as an hair
> Betid to any creature in the vessel
> Which thou heard'st cry, which thou saw'st sink. Sit down;
> For thou must now know farther.[50]

Here Prospero reveals a sanitized sublime: what had seemed to be the awesome power of nature to destroy the strongest works of man is in fact merely a "direful spectacle," safeguarded by art in which not a hair of any "creature" has been hurt. Shakespeare puns on "farther" as an amalgam of "art" and "father," while also creating a continuum between the art of spectacle and the art of procreation: "my art" refers both to his magic

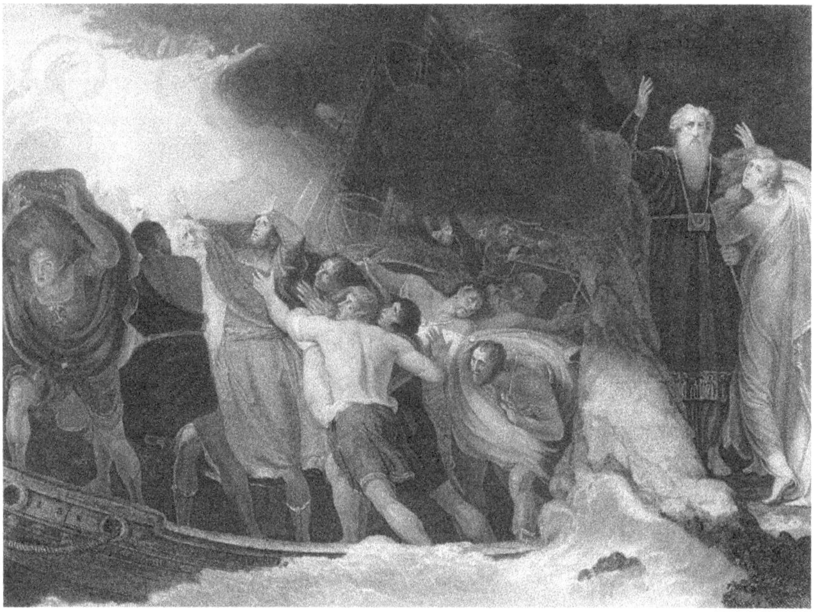

FIG. 3. *The Shipwreck,* Benjamin Smith, 1797, act 1, scene 1 of Shakespeare's *The Tempest.* (Library of Congress, Prints & Photographs Division, LC-DIG-pga-03317)

cloak and to Miranda herself. To "know farther" is to understand that her father is the keeper of an art form, and yet the ultimate source of the art (Ariel) as well as Prospero's intentions (to wed her with Ferdinand) remain unrevealed.

While there is much of interest in *The Tempest* in the relations between Prospero, Ariel, and Caliban, too often it is spectacle itself that takes the place of character-driven drama of the kind found in earlier works such as *Hamlet* or *Richard III*. This can be read in the stage directions, which call for increasingly extravagant and lavish depictions of thunder, mythical creatures, and transmogrifications. For all its atmospheric aspirations, Shakespeare wrote *The Tempest* for the *indoor* theater of Blackfriars Hall rather than the open-air setting of the Globe Theater.[51] Scholars debate whether Shakespeare himself wrote the stage directions or whether (as seems more likely) they were embellished by the scrivener Ralph Crane for publication in the First Folio. Most interesting is the suggestion that Crane worded the stage directions to reflect performances of *The Tempest* that he saw after Shakespeare's death.[52]

Regardless, the stage directions showcase an increasingly sophisticated theater where visual effects become the driving force of the audience's experience. Any contemporary viewer can likely attest to this: watching *The Tempest* is a singularly stirring and pleasant experience, where the easy dramatic unity of time, place, and action allows the visual fanfare to be perfectly enjoyed. *The Tempest* operates in the tradition of Renaissance masque in which the antimasque, the antimediation of disorderly, wild nature (Caliban), gives way to the masque proper of civilization and refinement. Spectacle itself is the mediating factor in this process for both the characters onstage and the creatures in the audience. Ferdinand, Miranda, Alonso, even Caliban, all are made to see a different world and then safely return to their previous positions with new insights, mea culpas, and emotional growth. In the Anthropocene, however, it seems unlikely that such harmonious resolutions will occur. Instead, humans are tasked with the responsibility of living with antimediation, to push back against the easy escapism of spectacle and sift through the difficult process of restoring and repairing ecosystems.

Local responsibilities increasingly operate within a more all-encompassing, world-scale design and thinking. While poioumena present a certain commentary on world-making, there is also an extensive history of planetary media that views the Earth as a single canvas to be ordered and arranged in aesthetic efficiency. Works from scholars such as Ursula Heise, Kelly Oliver, and Tiago de Luca take on questions of sensation, the sublime, and global consensus.[53] De Luca points out the paradox that "while planetary thinking is of the utmost urgency, the discourses that could provide models for this endeavour are implicated in troubling histories and lineages."[54] Iconic photos of the planet taken from space such as *Earthrise* or *The Blue Marble* reframe terrestrial position as that of an island floating in the void. At the same time, this abstracted and holistic thinking goes against much of what critical theory has sought to dismantle in its pursuit of relativized truths and socially constructed views of value systems. This book, while deconstructive at times, is also interested in looking at examples of world-scale thinking and design projects to better understand the history of globalization and the ways in which media have shaped ideas of the planet over time.

Modern meteorological technology continues to pursue the fantasy of Prospero's weather control. In 1922, British meteorologist Lewis Fry Richardson published his *Weather Prediction by Numerical Process.*

Richardson sets out in this work to present a model of forecasting that is not based on historical weather observation but can work from current conditions to predict future atmospheric changes. After much consideration of the calculus and equations behind weather prediction, Richardson indulges in a fantasy of a weather "forecast-factory":

> Imagine a large hall like a theatre. . . . The walls of this chamber are painted to form a map of the globe. The ceiling represents the north polar regions, England is in the gallery, the tropics in the upper circle, Australia on the dress circle and the Antarctic in the pit. A myriad computers are at work upon the weather of the part of the map where each sits. . . . From the floor of the pit a tall pillar rises to half the height of the hall. It carries a large pulpit on its top. In this sits the man in charge of the whole theatre; he is surrounded by several assistants and messengers. One of his duties is to maintain a uniform speed of progress in all parts of the globe. In this respect he is like the conductor of an orchestra in which the instruments are slide-rules and calculating machines.[55]

In Richardson's formulation, weather prediction is not just a calculation of the complicated systems of nature, but a delight in the orchestration of environment (see fig. 4). Richardson would go on to consider how his forecasting models might also be applied to the "storm" of war to equip leaders with tactical advantages.

At the core of Richardson's fantasy is a faith in understanding planetary systems. Yet the multiple autopoietic systems that humans set in motion prove canny and resistant to control. For this reason, as Joanna Page argues, "art may play a more powerful role in engendering a genuinely planetary perspective when it pays attention to those forces we cannot compel. Gesturing towards the *limits* of human agency with regard to the Earth may ultimately be a more effective way of challenging anthropocentrism."[56] Page suggests that more than careful control and precise measurement, artistic and critical methodologies may find more success by *letting go*.

Today the theater of global surveillance and forecasting is more sophisticated and exhaustive than ever before. There is more data, more knowledge, more understanding of the changes occurring to the earth. These changes can be seen, measured, and forecasted. The question remains, however: To what extent will this technological mediation change

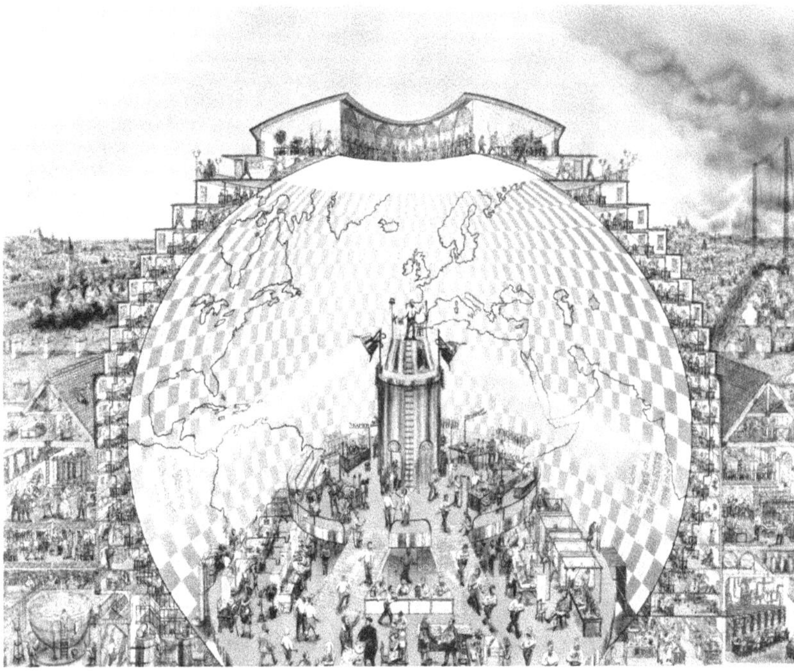

FIG. 4. *Weather Forecasting Factory*, based on *Weather Prediction by Numerical Process* by L. F. Richardson, Cambridge University Press, 1922, and on advice from Professor John Byrne, Trinity College Dublin. (© Stephen Conlin 1986, all rights reserved)

relationships to resource consumption, labor exploitation, habitat destruction, and species extinctions? In his exhaustive study of climate modeling, *A Vast Machine*, Paul Edwards situates Richardson within a larger trajectory of "voluntarist internationalism" and "quasi-obligatory globalism."[57] Richardson, in Edward's analysis, developed the forecast factory to contend with the "burden of calculation."[58] At the same time, Richardson's elaborate numerical techniques led to often very inaccurate weather predictions.

Only with the emergence of computer technology have meteorologists returned to a statistical model of complex computations and formulas. Edwards's comment on Richardson's fantasy is instructive: "Richardson's striking metaphors of calculation as factory, theater, church, and orchestra reach to the heart of computing as a coordinated human activity that harmonizes machines, equations, people, data, and communication systems in a frenetic ballet of numerical transformation. At the same time, they stand in stark contrast to today's dominant metaphors of computation, which are mostly individual: the brain, memory, neurons, intelligence. Richardson's forecast-factory remains a better description of the practical reality of computing. The limits of computer power, even today, stem from these human and material dimensions."[59] Challenges to global environmental control also, importantly, come from cooperative human and material resistance.

As computation networks turn away from individual humans toward dispersed cloud computing, data algorithms, machine learning, and artificial intelligence, it is more vital than ever to consider the *limits* of technological power. These boundaries are clear for the material demands, waste, and unexpected byproducts of technical intervention. However, the limits for human experience with technology are less well understood. While there are ways to measure addiction, isolation, and other negative effects of media immersion, it is not so easy to take a scientific approach to understanding the consequences of a world increasingly interacting through digital mediation. For this reason, the following chapters pursue a comparative approach that turns books, art, films, and software toward one another to reflect and refract the costs and benefits of media consumption. But, first, a bit of history.

2

Natural Historicities

> The one duty we owe to history is to rewrite it.
>
> —Oscar Wilde, *The Critic as Artist*

Christian Nyampeta's 2018 film, *Sometimes It Was Beautiful,* explores the intersection of media, empire, and exploitation. Focusing on the work of Swedish filmmaker Sven Nykvist in the Congo, the film juxtaposes historical violence, filmic violence, and the ongoing exploitation of Congolese nature and labor. It is a film about filmmaking: intercutting scenes from the Swedish movies with an imagined discussion between Nykvist and an audience of global leaders such as Wole Soyinka, Leela Gandhi, Robert Mugabe, Yasser Arafat, and others sitting in a movie theater. The film links the practice of colonial filmmaking to the ongoing ruptures created by imperial incursion. As one of the speakers puts it, "The damage of this film is the footprint it has left on history."[1]

After the dissection of Nykvist's intentions, the film shows the exploitation of the Congo rainforest and ties this extraction explicitly to the pulping of paper and by extension the production of knowledge in the publishing industry. Close-up shots of trees being harvested en masse give way to the chemical treatments and byproducts of paper processing plants. Clean rolls of white paper give way to a hygienic and organized bookshop where a woman peruses the shelves, perhaps looking for her own history. In the next scene, a man surveys a museum storeroom of stolen art objects, confronting the imperial appropriation of culture. While there is a certain didacticism to this approach, the film is far from a lecture. Instead, any true representation of colonial history or decolonization is eschewed in favor of abstract scenes of dancers in relief against a glowing red background or the sheer energy of a dirt bike rider circling through clouds of dust in slow motion.

The recent installation of this film in the atrium of the Guggenheim Museum in New York City situated the work alongside other environmental films such as Enoh Lienemann's *Dirty Waters* as well as drawings, written excerpts, and other works by Nyampeta on the walls. The suspended screen allowed the viewer to circle the film while walking along the famous ramp system designed by Frank Lloyd Wright in homage to the endless modulation of the car parking garage. Wright wanted museum goers to feel gravity as they viewed an art exhibit, a constant reminder of the pull of time and space in the heterotopic world of art.

This system has interesting effects when viewing a film: while one can sit on a bench, the experience is enhanced when detached from a static position. Not unlike the images of the motocross rider doing donuts, viewers can spin their wheels in constant renegotiation of viewing angle and attention to create a more interactive and multidirectional experience. Orbiting a suspended screen is a fitting metaphor for life in a Mediacene. Humans gravitate toward the play of light and energy, while the proliferation of media products vie for attention. A film like Nyampeta's is a poignant reminder that the history of media is imbricated with colonial history. As Alexander Galloway puts it, "History may be understood as the *organization of mediation*. But it is an organization of mediation that, at the same time, invents mediation."[2] Mediation must confront this intersection of history, aesthetics, and politics.

Golden Spikes and Diamond Jubilees

Any starting point for the Anthropocene proves to be somewhat arbitrary, yet critics offer no shortage of golden spikes, turning points, and points of no return. While there is no one place to point the finger, three influences are consistently identified: industrial revolution, colonialist expansion, and technological innovation. All of these factors are present in the United Kingdom's global dominance in the late nineteenth and early twentieth centuries, and indeed some have gone so far as to consider the current epoch an "Anglocene."[3] While critics have discussed the impact of the steam engine, indigenous genocide, and plantation slavery as key factors, less attention has been given to the proliferation of media networks alongside imperial expansion.

This chapter analyzes historical mediation more broadly as well as the role of media in global imperial expansion in the long twentieth century. Leo Marx summarizes a postrevolutionary view of industrial history succinctly as "a record of the steady, cumulative, continuous expansion of human knowledge of—and power over—nature."[4] As a result of this process, Dipesh Chakrabarty points out, the distinction between human history and natural history starts to break down. Most seem to assume that natural history becomes subsumed by human history, but in fact the opposite may be true: natural history, reeling from industrial activity, looks poised to increasingly dominate the trajectory of civilizations in the coming decades. Critics have given less attention to the similar fallout occurring in the realm of aesthetics. At the height of imperial and industrial activity, art shuns nature to pursue new autonomies of form. At the same time, the political failings of fascism, world war, and imperial decline return the arts to the social realm with a newly invigorated environmental focus.

Arguably the definitive critic of Victorian England, Oscar Wilde had little regard for history. To rewrite history was to invite the arts and creativity back into a realm that purported to deal only in facts. Yet he had, perhaps, even less regard for nature. He considered her—with casual misogyny—a foppish muse, monotonous, crude. One could rewrite history. But what could one really do with nature? Much better to invent. To create. In "The Decay of Lying," he stages a debate between Vivian and Cyril, with the former arguing that art is a gallant attempt to "teach Nature her proper place."[5] Wilde found no comfort or relief in the natural landscape, preferring instead the built environment and the intricacies of furniture design. To be modern was to deck oneself and one's immediate surroundings with the luxurious accoutrements of imagination to affirm the central place of the cultured ego. Yet, to accomplish this feat, to have all the furs and gold leaf and hand-carved wood, required a sophisticated control and exploitation of natural materials and labor allowed by the industrial and imperial heights of Great Britain.

Nature here is somewhat of a straw woman for Wilde's true target of realism and the scientific spirit of the age. Facts, it would seem, are the enemy. Charles Dickens also had something to say about facts in his iconic industrial-era critique *Hard Times,* where the fact-loving Gradgrind grinds out fact-packed graduates. Over the course of its installments, the novel showcases the shortcomings of a life built on facts. Gradgrind sees the downfall of his utilitarian ways when one of his formerly ground-out graduates, Bitzer, arrests his son Tom. Asked whether he has a heart, Bitzer cites the fact of William Harvey's blood circulation, making Gradgrind's entreaties to compassion, pity, and emotion seem merely nonsensical.[6] As an alternative to this dreary life of factuality, Dickens juxtaposes the circus, the protospectacle par excellence that celebrates the misfit and the outcast.

Unlike the realism and naturalism filling pages of post-Darwinian prose, Dickens's picaresque, ironic, satirical style presents a realism of a different stripe: what you see is what you get. Characters carry names that enforce their personality; their physical appearances match their inner natures. There is a realistic veneer, but it is always subservient to the plot, to the author's machinations, and to the moral resolutions of his tales. This style resulted from the subscription model of serialized literature that Dickens pioneered. The reader needed to grasp a story immediately, with minimal recourse to prior installments or recaps.

As media products, *Household Words* and *All the Year Round* contain the logic of the soap opera, the sitcom, the movie franchise, the video game series, the podcast, and every other serialized media form. Deliver a fully realized world into the domestic sphere, make it light enough to take anywhere, and keep it going fifty-two weeks of the year. Further, Dickens embraced the role of literary celebrity, setting up public readings and tours abroad to facilitate media as an event as well as a product. More than simply an additional revenue stream, this public element helped to bridge the gap between the potentially alienating world of private media consumption and the public recognition of shared interests and values.

Leaving aside the question of realism as aesthetic style, it is worth considering the influence of such a media stream on *reality*. Wilde's famous provocation that life imitates art puts a premium on stylistic invention for determining the future development of civilization. Environmental efforts, in comparison, can only go so far. Vivian: "We try to improve the conditions of the race by means of good air, free sunlight, wholesome water, and hideous bare buildings for the better housing of the lower orders. But these things merely produce health, they do not produce beauty."[7] Beauty, for Wilde, is the all-important resource of modern life. By insisting that nature is *not* beautiful, contrary to prevailing theories of environmental aesthetics, Wilde sets up a new search for the beautiful in that which is created by humans.

But what exactly does Wilde mean here by "beauty"? Is it merely an aesthetic taste, a visible façade? Who determines what is beautiful and what is not? Vivian points to the pre-Raphaelite painters for influencing the standards of beauty of the day. Works such as Edward Burne-

Jones's *The Golden Stairs* and *The Beguiling of Merlin* prompt imitation from viewers. If one is to take the pre-Raphaelites at their word, one quickly arrives back to a discussion of nature. The journal *The Germ,* established in 1850 by Dante Gabriel Rosetti, William Michael Rosetti, William Holman Hunt, and Everett Millais, aimed to promote the "thoughts of Artists, upon Nature as evolved in Art." What this means, exactly, is unclear, as further elaboration appeals to an equally vague "rigid adherence to the simplicity of Nature."[8] It appears Wilde has been caught in a bind of sorts. If "nature" is of no interest to art, then why does he celebrate the stylistic influence of a band of artists who hold nature and a natural approach in such high esteem? Of course, contradiction is central to Wilde's own literary art, but, logical loops aside, what remains more dangerous in his appeal to the pre-Raphaelites is a covert endorsement of "natural" or "naturalized" standards of beauty. The wan, fair-haired, pale-eyed procession of young women that descend Burne-Jones's spiral staircase form a rhythmic repetition of a "type" or idealized female form.

This obsession with a certain portrayal of a more "natural" femininity, at the hand of a painterly "brotherhood," came under scrutiny from poet and sister to aforementioned brothers, Christina Rossetti. Rossetti is best known for her long poem "Goblin Market," which portrays a feminine, sapphic resistance to the orgiastic masculinity of global market economies. But another of her poems, "In the Artist's Studio," neatly critiques the male artist's gaze. The opening line of the Petrarchan sonnet, "One face looks out from all his canvases," presents a painter's reductive vision of his female model. Artistic ineptitude may seem benign enough; however, the poem takes a more sinister turn in the sestet as "he feeds upon her face by day and night," suggesting an act of consumption rather than creation. As for the art spawned by this obsession, "every canvas means / The same one meaning, neither more or less."[9] Instead of growth and potential, the artist's studio becomes a static place of repetition. The focus on an empty notion of beauty limits the artist's potential for *expression,* what Wilde appeals to as the "energy of life."[10]

No longer some mere surface quality of appearance, to see beauty becomes a recognition: an understanding of how the viewer's being is entangled in some exterior object or environment. Consider, for example, Monet's series of paintings of Waterloo Bridge (see fig. 5). Composed over several winters at the turn of the century, Monet painted the bridge over forty times. What was it about this scene that attracted Monet,

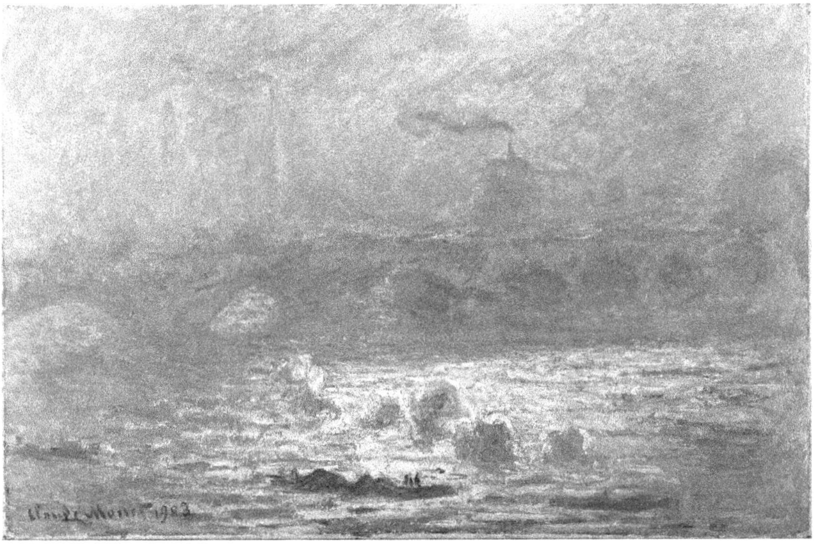

FIG. 5. *Waterloo Bridge,* sunlight effect, Claude Monet, 1903. (Denver Art Museum; funds from Helen Dill bequest, 1935.15; photography courtesy Denver Art Museum)

year after year, as he sat on his balcony at the Savoy Hotel? Perhaps it was nothing, merely an arbitrary point of reference for a series of practice canvases. Monet, the Frenchman, may have been sensitive to the bridge's commemoration of the Battle of Waterloo, where the Napoleonic era came to its kerplunk end. Napoleon's downfall—pun intended—was overextension. At Waterloo he met the bridge of forces between England and the Netherlands. Was Monet simply trolling Napoleon by painting this bridge over and over again? Or had Monet read Thomas Hood's 1844 poem "The Bridge of Sighs," eulogizing the death by suicide of a homeless woman, not unlike the fallen sisters of Rosetti's "Goblin Market"? Although the world has spurned this unfortunate, she is accepted by the natural elements, as "the wave constantly / Drips from her clothing," and her lips ooze "clammily." The anonymous woman, Hood argues, was "mad from life's history."[11] Like the mud that builds up around the bridge pillars, history accumulates, becomes heavy, leaves people stuck.

Shortly after Monet's own death the bridge would undergo plans to be rebuilt, although actual work was delayed until World War II. Largely completed by women, newly called to the workforce, the bridge of sighs became "the Ladies' Bridge." Stones from the original were sent around the commonwealth, lodging in disperse locations such as Canberra, Wellington, and Mutare—a fitting image for the end of the Anglocene, ruins of its former glory dispersed throughout an increasingly independent commonwealth of nations.

Regardless of Monet's motive, what seems significant for an analysis based in environmental aesthetics in his series of bridge paintings is the juxtaposition of fixity and movement. While the bridge remains static, it is surrounded by the movement of the water as well as the skies: the smokestacks on the far shore emphasize the passing of time and the rhythms of production as do the pedestrians and horse carts bustling over the pass. Monet has no interest in mastering or controlling the image. Instead, he allows the landscape to fluctuate in painting after painting to reflect current conditions, atmospheric resonance, and perhaps his own mood or emotions. He collects data. The shift from the male gaze of the pre-Raphaelite's could not be more complete. The bridge is no projection of Monet's hopes or desires. The landscape asserts something more humble: a painter in solidarity with environment. The distinction between art and nature begins to break down, and an interrogation of the material world begins in earnest.

At the risk of cliché, art is also a bridge between painter, viewer, and world. The aesthetic—in the context of *sensation*—allows humans to explore, to arrange, and to learn from their material surroundings. When Wilde attacks nature he cuts off this bridge. A hermetic turn ensues, one best exemplified by Joris-Karl Huysmans's *À Rebours.* Translated as "against nature," the more literal "reverse" or "inversion," a sense of going inward speaks to the escapism of the modern age. Des Esseintes's foray into a life of luxurious adornment failed, but might he have been successful with the aid of a big-screen TV, a satellite dish, or, better yet, unlimited high-speed WiFi, a VR headset?

Wilde's attacks on nature are more political than those made by Des Esseintes. The invocation of nature by moralists of his age to enforce heteronormative sexuality no doubt rubbed him the wrong way, so much so that he was willing to put success and social acceptance on the line in another type of performance—a grand fin-de-siècle courtroom spectacle—that would eventually land him in the decidedly harsh reality of work prison. Perhaps it is not so much the natural world that Wilde spurns but a brand of natural sciences, natural histories that raise empiricism to an ultimate position of authority. Only after a century of staggering insights into the natural world could this blasé attitude be so provocative.

An Instauration of Natural Histories

Rather than the contemporary geologists of the day such as Charles Lyell and his successors, Wilde preferred the natural history of Pliny the Elder. Pliny's introduction to the reader—most immediately, the emperor Titus, but presumably also anyone interested—details the scope and style of the work, explaining that the following books speak of "the Nature of all Things in this World" in "barren Terms, without any Show of Phrases: and what I have noted concern the commonest Points thereof, so that I am to deliver the Matter either in rustic, or foreign, nay, even barbarous Language."[12] Pliny pursues a reliable and straightforward style to mimic the honesty of the natural world itself. At the same time, Pliny was drawn to the mysteries of nature, and here an imaginative streak where facts take a backseat emerges to explain—or at least grapple with—phenomena not easily understood by the observer.

It was this ceaseless curiosity that led Pliny to his own death by asphyxiation at the volcano Vesuvius, where he stayed to help rescue victims after the eruption in 79 AD that destroyed and fossilized Pompeii. Pliny, "spurred on, no doubt by his scientific curiosity . . . lingered to observe the volcano at close quarters. Landing at Stabiae from a small boat he was overcome by the sulphurous fumes and died on 24 August 79."[13] Pliny's writings launched the field of natural history, and with the rise of modern science and the advent of new technologies, humans have a greater understanding of the natural world than ever before. And yet the unexpected volcanic eruption of nature remains a persistent threat.

Pliny's writings were of great interest to Francis Bacon, even as he was wary of the pagan and mystical elements. Bacon's *Instauratio Magna* (*The Great Instauration*) provides a comprehensive reappraisal of many fields of scientific inquiry. In some ways, the Anthropocene is a result of this great instauration, allowing human populations to grow in orders of magnitude. The technosphere is also a reflection of all that has been *instilled* into the environment as a result of this knowledge: objects, pollutants, chemicals, emissions, and other matter that has shifted the geological makeup of the planet. If Bacon's new organon presented a significant way forward for interpreting nature, it came at the expense of a more open, imaginative, and associative relationship with the natural world.

In *Sylva Sylvarum,* Bacon distinguishes his own approach from Greek thinkers such as Pythagoras and Plato. While Bacon rails against superstition, an extended look at his critique reveals that the imaginative view of earth's interconnectedness might find new relevance:

> The philosophy of Pythagoras, which was full of superstition, did first plant a monstrous imagination . . . that the world was one entire perfect living creature . . . the ebbing and flowing of the sea was the respiration of the world, drawing in water as breath, and putting it forth again . . . the sense and the effects of any one part of the body instantly make a transcursion throughout the whole body: so that by this they did insinuate, that no distance of place, nor want of indisposition of matter, could hinder magical operations; but that, for example, we might here in Europe have sense and feeling of that which was done in China; and likewise we might work any effect without and against matter; and this not holpen by the co-operation of angels or spirits, but only by the unity and harmony of nature.[14]

Such proclamations of the world as a living, breathing biosphere no longer seem outlandish in contemporary discourses surrounding the ever-connected webs and networks of globalization. Emissions from China *do* affect Europe, and vice versa. Moreover, this view returns not simply through superstition or imagination, although certainly these elements persist in Gaia theory; instead, it is scientific advancements made by geologists such as Charles Lyell and biologists such as Charles Darwin that have led the way toward an articulation of ecology as a system of interconnected life forms.

Lyell is not dismissive of previous explanations of natural phenomena, but instead sees belief systems as attempts to understand and navigate major geologic changes. Just as past events in human history have shaped contemporary society, so have past events in natural history shaped the contemporary environment. Lyell is quick to note the economic stakes of the new discipline, most notably in the discovery of fuel to power modern nations.[15] Geology and extraction go together: geologists point the way for mining operations, and as workers dig deeper into the earth, geologists are there to observe new phenomena and explain the different layers of sediment, the fossils, and the aberrations contained within the earth's crust. For Lyell, geology is tied to physical science just as history is tied to moral science. The geologist and the historian are aligned by a common pursuit into the truth of the past, one that must be drawn with high standards and precise measures. He takes care to distance geology from cosmogony: he positions himself as a simple reader of the earth, of its layers and strata, rather than a keeper of truths or ultimate wisdom about the world's origin and humanity's relationship therein.

Principles of Geology makes an exhaustive survey of philosophers and naturalists throughout time and from different corners of the world. Lyell positions the Scottish geologist James Hutton as his predecessor in modern geology for his theories of the long, slow, and continual processes that have formed the physiognomy of earth. With Hutton, scientists moved beyond plutonist and neptunist debates to confront the deep time of natural history: "The imagination was first fatigued and overpowered by endeavouring to conceive the immensity of time required for the annihilation of whole continents by so insensible a process. . . . Such views of the immensity of past time, like those unfolded by the Newtonian philosophy in regard to space, were too vast to awaken ideas of sublimity unmixed with a painful sense of our incapacity to conceive a plan of such infinite extent."[16] The sublime, as outlined by Kant, Burke, and others, occurs primarily through phenomena in nature that overwhelm the observer through sheer scale. Here, however, it is not simply physical size that dwarfs the human, but time itself.

What would it mean to reconsider the sublime to be not merely a spatial experience but to encompass temporal awareness as well? This sensation is becoming more common in the wake of modern geological insights. Previously, people could stand separate from the sublime to find security and view it with a sense of preeminence that asserted civilization's domination of the natural world. Now, however, nature is no longer nature alone, but has been augmented, made strange, and more threatening from human activity. While Lyell asserts that the observer in the Victorian era can find solace in the notion that much of geology is a slow and relatively peaceful process, contemporary observers have no such consolation as they face a dizzying acceleration.

One thinker who holds some lessons for grappling with a geologic sublime that is temporal as well as spatial is Baruch Spinoza, whose *On the Improvement of the Understanding* presents a parallel take on knowledge to the one provided by Francis Bacon. Rather than dismiss attempts to see the world as a holistic, living organism, Spinoza invites a process whereby the individual can expand into the world through knowledge:

"The mind apprehends itself better in proportion as it understands a greater number of natural objects . . . we shall infer from these ideas other ideas, which will in turn be connected with others, and thus our instruments for proceeding with our investigation will increase."[17] Spinoza details an expansive process of mind, one in which the individual radiates out into the world in greater chains of knowledge and understanding.

Perhaps a similar process is available to the mind that confronts the physical reality of deep time: an approach that embraces slow accretion rather than attempting to hold billions of years of past time in the mind for an instant. While this process is well and good for individual growth, this accumulative process of gathering natural objects for the betterment of civilization is also part of the logic that seeks to justify imperial incursions into foreign territory. To extend this discussion to the workings of institutions and nations, the following section considers how the conquest of nature and the media distortion of nature's image went hand in hand under the auspices of the United Kingdom's Imperial Institute.

Mediating Imperial Decline at the Empire Story-Land

Lyell published the first edition of his *Principles of Geology* in 1830. He updated the work constantly, and by the eleventh edition of 1872—the last published in his lifetime—there were numerous changes and updates to the field. Foremost among these was the work of Charles Darwin, whose 1859 work *On the Origin of Species* drew heavily on Lyell's geology even while his articulation of evolution surpassed Lyell's geographical theory for the distribution of species. The science of Lyell and Darwin benefited from the expansion of the British Empire under the steady rule of Queen Victoria from 1837 to 1901.

As the United Kingdom made its way around the globe, human and natural histories intersected in new ways. In 1888, the Imperial Institute was established by royal charter in order to "promote the utilization of the commercial and industrial resources of the Empire by arranging comprehensive exhibitions of natural products . . . and providing for the collection and dissemination of scientific, technical and commercial information relating to them."[18] Throughout the publication of its yearbook and bulletins, the institute presented a "Statistical View of the British Empire," detailing close to a billion pounds sterling in imports and exports in livestock, cheese, cotton, flax, jute, silk, oil, coal, and numerous other natural products. The volumes give different histories of the various British "properties" throughout the empire, but the real history seems to unfold in the various charts, graphs, and tables that track the emergence of a modern global system of capitalism.

If the Imperial Institute served as the historical register of the natural world being bought and sold, its involvement in the British Empire Exhibition of 1924–25 signaled an initial attempt to stage a cultural history of Empire. The exhibition set up quarters at Wembley Park outside of London, presenting an open air studio set of the British Commonwealth. A handbook published for the exhibition outlines its attractions and scope, aiming to recreate the entire world as a leisure park. The exhibition would take stock of the empire's resources to organize efforts and increase production, finding new "possibilities of growing more grain and wool, more sugar and more cotton, more timber and more fruit; of mining a larger quantity of metals; of adding in every way imaginable to our production of commodities which meet the world's needs."[19] To this end, the exhibition was also the site for the first "World Power Conference," which assembled scientists, engineers, industrialists, and manufacturers from around the globe to discuss fuel sources and production methods.

Alongside exhibitions of minerals, timber, and fishing industries, the British Empire Exhibition included music and entertainment to provide "amusement and relaxation."[20] Yet a closer look at events such as the "Pageant of Empire" reveal these amusements to also be vital forms of promoting a particular narrative about Britain's imperial rule. An internal memo explains that the "underlying motive of this Pageant of Empire is to give a representation of the Empire, Imperial in its characteristics, and more especially showing the close connection of the history of the service and the growth of the Overseas Empire."[21] Here a particular slippage occurs between the official history of British imperial conquest and the "representation" of this same project of colonization as a heroic project. A bit of a kerfuffle occurred in the controlling of this image when the Prince of Wales sent out a letter that invited representatives of the colonies to assist in the pageant. The memo clarifies: "There was no idea of asking the Colonies (or to put up the money) to bring contingents of natives from any part of the Empire. Whatever representation of any particular part of the Empire was considered necessary, the performers would be drawn mainly from residents in the United Kingdom, the idea being to obtain advice of the Representatives of the particular part of the Empire concerned as to the accuracy of the representation."[22] Instead of allowing colonial subjects to represent themselves directly, the memo specifies that this will be a pageant firmly controlled by the imperial center. In terms of story and presentation, the pageant confirms this imperial bias.

The program for part 3 of the pageant—focused on New Zealand and Australia—presents a romanticized account of Captain Cook, war with the Māori, and the emergence of "King Wool" as bringer of prosperity and peace. The native Māori people are presented as noble but ultimately outmatched by the "superior," "tactful," and "courageous" settlers. The end of the New Zealand pageant dramatizes a new solidarity in the face of the global threat of World War I: "The Union Jack is unfurled, and men of all kinds and classes, including the Maoris, fall into rank together and march off to martial music in answer to the Call of the Empire."[23] The pageant performs a specific view of historicity: one in line with a philosophical tradition that sees civilization as a progressive force that may suffer setbacks but ultimately tends toward unification, prosperity, and liberty. Here is a highly selective history, told by a carefully selected subset of the population, within a curated exhibition of stories about empire.

A different perspective on the British Empire Exhibition comes from Virginia Woolf, whose essay "Thunder at Wembley" satirizes the event to show how fragile and fleeting Britain's world dominance might be. Woolf portrays the exhibition as a glorified flea market where everything is on sale for six and eight pence. Against the "ferro-concrete" artificiality of Wembley Park, however, nature stands out all the more. The products of empire appear inert and lifeless, but the people who peruse the items come to life in new ways hitherto unseen. While most of the exhibitiongoers seem narcotized, walking around in a stupor, Woolf also suggests that a revolutionary new perspective might emerge through the rediscovery of the natural world: "You look up and discover a whole chestnut tree with its blossoms standing; you look down and see ordinary grass scattered with petals, harbouring insects, sprinkled with stray wild flowers."[24] To Woolf, the exhibition does not signal the realization of humanity's historical destiny. If anything, it becomes a scene of paralysis, an artificial dead end for cultural development.

The remainder of the essay presents the destruction of the exhibition in dramatic fashion. Failing to control the sky itself, the marching band of empire gets blown away by a violent storm. What had seemed so permanent and lasting before is unmade in a few moments: "Pagodas are dissolving in dust. Ferro-concrete is fallible. Colonies are perishing and dispersing in spray of inconceivable beauty and terror which some malignant power illuminates."[25] Woolf inverts the official narrative of empire as triumph over the natural world through a vision of humanity's susceptibility and fragility. People do their best to run for cover, some even cling to a butter statue of the Prince of Wales, but it is all in vain as nature reasserts itself.

While Woolf questions the success of the British Empire Exhibition, the mode of historical storytelling it enacted continued to influence the United Kingdom. By the 1930s, the Imperial Institute had rebranded itself as "An Empire Story-Land" whose galleries, cinema, and film library comprised an "Art of Visual Instruction." While the Imperial Institute began as a statistical endeavor, within fifty years the numbers and accounting of empire became less important than the stories behind these numbers, and the stories themselves were curated to promote common experiences and emotions that would allow viewers to connect to an imperial sentiment that appeared historically destined. As H. A. F. Lindsay, director of the institute, argues, "The story of an economic product can be told in such a way that it remains a more or less permanent memory. When it is associated with a particular country, then a double purpose is served. The geographical and economic importance of the product are alike stressed, for the product suggests the country of origin and the country the product. . . . The attention of the visitor is not merely arrested but held, not merely held but intrigued, and so intrigued that, like Oliver Twist, he asks for more."[26] Lindsay suggests that the histories of people, places, and products are inextricably linked—not only in a natural sense, but also by the stories that are told about these processes. Here a literary element enters: To what extent do stories mediate the forces of history that have otherwise been understood as occurring naturally, guided by a world spirit, or environmentally determined? Citing Dickens, no less, he identifies not just an appetite for the raw products of the empire, the sugar and spices, but indeed an appetite for stories of empire, an appetite for content and media.

At its height, the Imperial Institute was a truly global media operation, sending and receiving films, photographs, pamphlets, slides, dioramas, and other educational materials throughout the vast network of British dominions and colonies. Among other collectibles, picture cards were sold as information items at a price of 6 pence a series. Packets ranged from *The Manufacture of Tea in India* to *The Rubber Industry of Malaya* and *The Clove Industry of Zanzibar.* The institute's film catalog in 1939 stood at hundreds of films ranging in duration from seven minutes to over an hour. Titles such as *Highways of the Empire, Animals of the Empire, Transfer of Power, Zulu Pomp, Story of South African Steel, Romantic India, Tea-time Topics, Jungle Gods, Happy Uganda, Presta Gold Mine, The Magic Carpet, Plantation People,* and many others attest to the range not only of subject matters and geographical locations, but also to multiple modes of storytelling within the different films.[27] While these were all theatrical releases, the institute also engaged in extensive marketing and library consolidation to increase the reach and viewership of the empire films (see fig. 6).

Many of the films in the collection were what Lindsay called "Background Films" meant to give a sense of foreign territories in the empire. Lindsay explains in relation to the film "Romantic India": "It does succeed in giving an impression of the vast ranges of climate and soil, agriculture, industries and markets, races and character and life, which make up the India which we know and love, so far, at least, as a twenty-minute film can suffice for the purpose . . . 'background films' . . . are, or try to be, true to nature or true to human nature . . . they must describe life as it really is and not mere lifelikeness."[28] Film in this analysis becomes more vivid than mere description. Indeed, it holds the potential to be more real than reality. Through its artful arrangement, the viewers may be led to insights that might otherwise elude them even if they were to physically visit India.

The exhibition arm of the institute was complemented by a scientific arm that was hard at work on ways to maximize materials and applications from numerous foreign resources, an effort that increased as the United Kingdom geared up for a return to war. The day World War II broke out, fourteen staff members transferred to the Ministry of Economic Warfare.[29] An article in the February 15, 1941, *Times Educational Supplement,* "The Imperial Institute in War Time," details how "laboratory tests were applied to products of such widely differing character and origin as resin from Ceylon, cinnamon bark oil from the Seychelles, lime oil from the Sudan, beans from Malaya, ginger from Nigeria, and boxboard made from Ceylon coconuts."[30] The institute's "List of Economic Products" totaled some 250 items from cashew nuts to dragon's blood, all of which came under reevaluation for the potential to assist in the war effort.

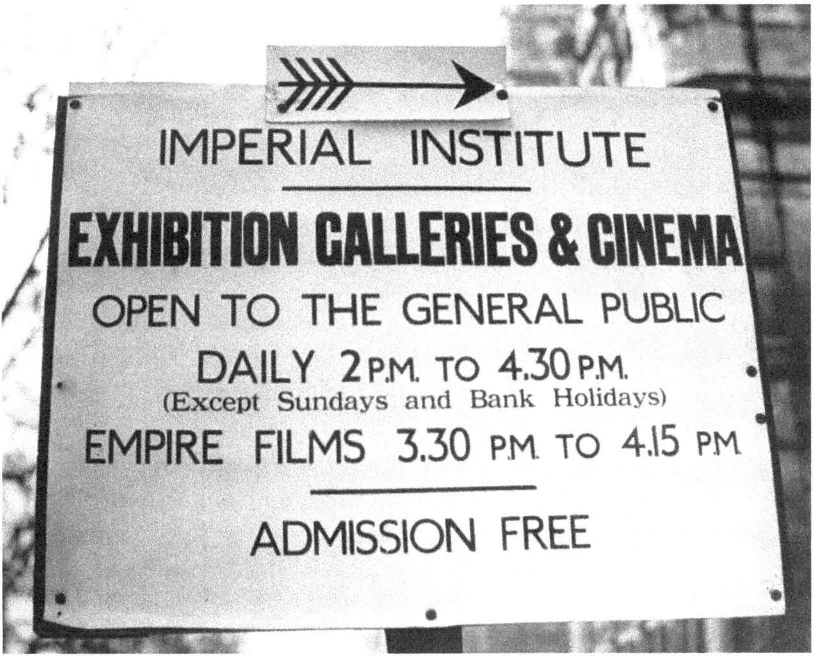

FIG. 6. Imperial Institute, exhibition galleries and cinema sign. (National Archives)

Combining art, science, and statistics, the Imperial Institute was one of the first information-driven data centers aspiring to operate at a global scale. Its progress, however, was halted alongside the contraction of the British Empire itself. In the postwar years, the institute would shift gears once again to be on the frontlines of mediating the decline of the empire. The future viability of the institute was debated in the Skelton Committee report of 1933, and immediately following World War II its continued viability was again under question. In 1946 the complete abolition of the institute was under consideration; however, the treasury advised against this drastic action because "the abolition of the Imperial Institute might be regarded as a symbol of the dissolution of the British Empire."[31] Rather than disband the institute, the memo advises breaking up the various functions performed (e.g., publicity, education, animal and mineral department, science lab) and continuing through other government bodies. Just as in the process of decolonization, here the instruments of empire are reconfigured within the imperial center to continue operating in a neocolonial fashion.

The major media shift occurring in the postwar era was to move from products to *people*. Again Lindsay: "It is only when we cease to be 'introverts' and begin to project our minds outwards towards the conditions and interests of other people that our education really begins . . . Every lesson in history or geography should have behind it some sort of emotional stimulus which will help it along and see it through."[32] Instead of simply selling products, now Lindsay reimagines the potentials of the institute to promote understanding and worldly spirit. Just as the film is projected onto a screen, so the mind is projected into another way of life, another world perspective.

To accomplish this shift, the exhibits and permanent courts were renovated and modernized (see fig. 7). A report on the "Exhibition Galleries" explains the desired changes: "The aim is, wherever possible, to replace the old-fashioned showcases with internally-lit displays built into superstructures designed to give unity and atmosphere to each national exhibition. The visitor will be invited *into* the country he is visiting."[33] Cinema attendance at the Imperial Institute was also greatly

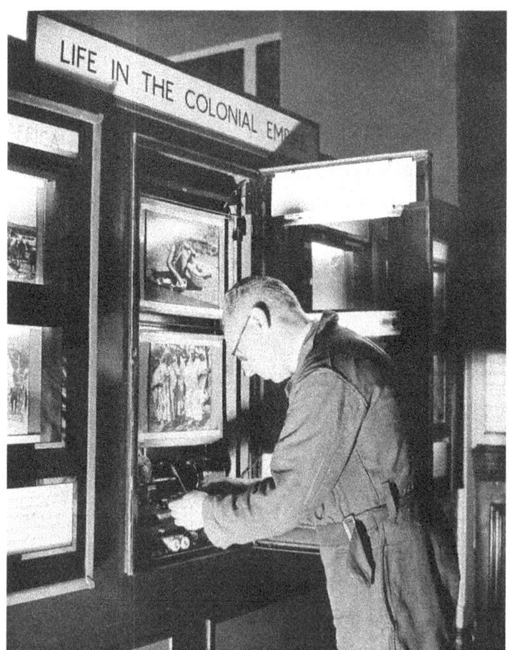

FIG. 7. Imperial Institute, life in the colonial empire. (National Archives)

increasing—tripling over the course of four years with the introduction of more color films. The move in both films and exhibits was to heighten the sense of immersion, to make the media tantalizingly real, a virtual reality. Old media was also robustly represented with the publication of thousands of books, maps, murals, leaflets, classroom dioramas, postcards, and picture sets.

The institute carried out this virtual mediation of populations alongside the management of geological resources. Speaking in 1952, Lord Ogmore defended the institute for its comprehensive approach to global terrain itself. "In conjunction with the Colonial Aerial Survey," he stated, "the work of this Institute in providing accurate data for explorers, prospectors and others concerned with minerals, cuts down the time of exploration not merely by months but by years."[34] The machinic technology of the airplane combines with media technologies for surveying and mapping as well as geological expertise to accelerate the exploitation of rare and valuable minerals. Thus, even as the empire contracts, its geological agenda for resource accumulation can continue apace. Further modernization came in 1958 when the name was changed to the Commonwealth Institute, in recognition of the shifting political landscape and the growing independence of former colonies. By the end of the twentieth century, the institute privatized into something of a transnational corporation itself, but this, too, went bankrupt, spiritually as well as economically, and the institute folded its assets into a nonprofit education trust.

The World of Mediation

The aggressive extraction of coal in the United Kingdom coupled with the invention of the steam engine became a major accelerator of climate change. Other nations followed this consumption-heavy model for driving global modernization. While European empires may have declined in the second half of the twentieth century, the exploitation of earth's resources has only increased, entering what political scientists term the "great acceleration." Many also consider the post-1945 era to be the moment of arrival for the Anthropocene as a global phenomenon. If the consequences of resource exhaustion are present throughout civilizations, going back perhaps to the *Epic of Gilgamesh* and the deforestation that would be the undoing of the Sumerian ruler, the post-1945 era presents environmental issues on a much larger and broader scale than ever before. Added to which, the detonation of nuclear weapons in Hiroshima and Nagasaki presented a sublime image alongside a material annihilation that encapsulated the ability to destroy all life on earth.

Nuclear waste and radiation are further measures of anthropogenic change. Beyond the sheer size of destruction, the advent of nuclear technology speaks to the ability to fundamentally alter the basic building blocks of nature. This was a future imagined by Francis Bacon in his work *New Atlantis,* which presents a utopian civilization that has managed to interpret, augment, and shape nature's scope toward the end of "Enlarging . . . the bounds of Humane Empire, to the Effecting of all things possible."[35] While Bacon imagines the benevolent and prosperous society that may emerge under the omnivorous exploration of nature, the advent of nuclear weapons and the growth of petrochemical industries speak to a far more dystopian outcome of the incursion into natural processes.

Hannah Arendt notes as much in her response to the unrestrained violence and destruction of nuclear warfare and World War II more broadly. By setting in motion new natural processes, Arendt argues that humans have "begun to act into nature as we used to act into history," with the result that "we have manifestly begun to carry our own unpredictability into that realm which we used to think of as ruled by inexorable law."[36] Arendt's formulation here may be one of the earliest and most succinct statements of the challenges facing modern civilization: despite the best intentions to keep art and nature separate, a large degree of uncertainty remains. Calls from geoengineers to act deeper and further into natural processes risk a similar unpredictability.

If Arendt identifies a shift in human action intersecting with natural processes, contemporary theorists consider a much larger scale of influence. Chakrabarty argues that "it is only very recently that the distinction between human and natural histories . . . has begun to collapse. For it is no longer a question simply of man having an interactive relation with nature. . . . Now it is being claimed that humans are a force of nature in the geological sense."[37] If this is indeed the case, as the overwhelming evidence of science would have it, what is perhaps needed is a new sense of the *historicity* of nature.

A basis for this critical inquiry can be found in the social theories of the Frankfurt school. Susan Buck-Morss explains how Theodor Adorno takes a stand against the philosophical synthesis of nature and history found in Heidegger's theory of "historicity." Adorno instead articulates nature and history as "dialectically opposed concepts, each of which provided a criticism of the other, and of the reality each was supposed to identify."[38] Walter Benjamin extends this dialectic to include discussions of technology, culminating in his unfinished *Arcades Project.* Employing geological metaphors, he writes, "Just as there are places in the stones of the Miocene or Eocene Age that bear the impression of huge monsters out of these geological epochs, so today the Passages lie in the great cities like caves containing fossils of an ur-animal presumed extinct: The consumers from the pre-imperial epoch of capitalism, the last dinosaurs of Europe."[39] The *Arcades Project* becomes an accumulative analysis, collecting data points of commodity capitalism's refashioning of the urban environment.

If philosophy pursues alternative forms of historical practice, literature also presents challenges to the assumptions and biases found in historicism. The nexus of imperialism, natural history, and environmental aesthetics finds purchase in César Aira's novella, *Un episodio en la vida del pintor viajero* (*An Episode in the Life of a Landscape Painter*), published in 2000 and translated into English in 2006. Aira presents a surrealist history of the German painter Johann Moritz Rugendas (1802–1858), a follower of Alexander Von Humboldt's (1769–1859) physiognomic method, which considers a conglomeration of biological, meteorological, and geological factors to approach the wholeness of nature. Rugendas seeks an alternative to this process in the pampas of Argentina where he suffers an episode (spoiler alert) consisting of getting struck twice by lightning and then dragged by his horse across the unforgiving plains until his own physiognomy is a tangled and rearranged mess.

With time, and morphine, Rugendas recovers and returns to painting. His interest shifts from picturesque landscapes to the violent and theatrical raids between the Indians and European settlers. Rugendas pushes the scientific preciousness of the physiognomic method to its limits to reveal a more immersive mediation with his surroundings. A passage toward the end of the novella explores this common connection within all forms of expression and creativity:

> The scenes followed one another in a certain order, but their traces on paper suggested other orders, which, in turn, affected the original scenes. . . . Over the course of the day, there was a progression—though it remained incomplete—towards unmediated knowledge. It is important

> to remember that their point of departure was a particularly laborious kind of mediation. Humboldt's procedure was, in fact, a system of mediations: physiognomic representation came between the artist and nature. Direct perception was eliminated by definition. And yet, at some point, the mediation had to give way, not so much by breaking down as by building up to the point where it became a world of its own, in whose signs it was possible to apprehend the world itself, in its primal nakedness. This is something that happens in everyday life, after all. When we strike up a conversation, we are often trying to work out what our interlocutor is thinking. . . . We come up against the words, and before we know it, we are already emerging on the other side, grappling with the thought of another mind. *Mutatis mutandis,* the same thing happens with a painter and the visible world. It was happening to Rugendas. What the world was saying was the world.[40]

The German painters seek to operate as a third term between the settlers and the indigenous. They are merely passing through, recording scenes as neutral observers. And yet their presence is far from innocuous. Rugendas is employed in the business of recording the scenes of colonialism and imperial conquest. He sells his paintings and etchings to a European market eager to consume exotic views of nature.

In the above passage, Rugendas, assisted by the anesthetic morphine, pushes beyond the physiognomic method of Von Humboldt, beyond the facts and data that so riled Dickens and Wilde, to make mediation its own world. He is trying to reclaim the *episode* itself, when the bolt of lightning bypassed his sensations, eschewed the aesthetic channels, and tapped directly into his nervous system. The traces of the scenes Rugendas observes provide an immediate historical record. But his observations also unlock parallel and alternate universes not unlike the bouncing of quantum particles. History becomes slippery, subjective, and ready for rearrangement.

The final scene of the novella brings Rugendas face-to-face with the Indians, furiously sketching and working his new process of mediation. He sacrifices his own physiognomy, veiled behind a mantilla, to push beyond the physiognomic process toward the lightning-bolt charge of connection. What, then, is the role of his art in this moment? Why does he need the third term to approach this community? Is this a moment of fellowship at all? Or is it simply another appropriation, an imperialist grab that takes and then leaves? Something different is happening in this scene. Rugendas maintains the importance of the aesthetic as a mediation not just between figure and environment but between person and person.

The task for literary engagements with natural historicity might not be so much to "rewrite" history, as Wilde would have it, but to "unwrite" history to allow for new forms of being and becoming to emerge. Chimamanda Ngozi Adichie's 2008 story "The Headstrong Historian" illustrates how different historicities come into conflict with one another. This short story follows three generations of an Igbo family, starting with the marriage of Nwamgba and Obierika and ending with the birth of Grace (Afamefuna), who becomes a scholar of Nigerian history. Obierika's family suffers from infertility, and each generation struggles to conceive. These personal anxieties occur in relation to larger, systemic changes to Nigeria wrought by British colonialism and Christian missionaries.

The teleological historicity of western European "being" and "becoming" threatens to supplant the Igbo conception of time as cyclical. At the same time, a connection between the first and third generations forms a solidarity against the "new ways" of the middle generation: "From the moment Nwamgba held her, the baby's bright eyes delightfully focused on her, she knew that it was the spirit of Obierika that had returned; odd, to have come in a girl, but who could predict the ways of the ancestors? Father O'Donnell baptized her Grace, but Nwamgba called her Afamefuna, 'My Name Will Not Be Lost.'"[41] Grace goes on to study history at the University College in Ibadan and, through her own research, eventually challenges the historical accounts of Nigeria presented by the British Empire. But what is perhaps most interesting in this final section is Adichie's choice to shift to a future tense for describing Grace's adult life: "It was Grace who, driving past Agueke on her way back, would become haunted by the image of a destroyed village and would go to London and to Paris and to Onicha, sifting through moldy files in archives, reimagining the lives and smells of her grandmother's world, for the book she would write called *Pacifying with Bullets: A Reclaimed History of Southern Nigeria*."[42] The story telescopes into Grace's future from the vantage of Nwamgba's present. The repetition of what Grace "would" go on to do presents a futurity where these events are already in the past.

In this historical future, Grace rediscovers the world of the first part of the story. By placing emphasis on what Grace "would" go on to do, Adichie creates a vision of the future that is equal parts guided by potential and predestination. The title of Grace's monograph also references the euphemism of "pacification" that Chinua Achebe critiques in his novel *Things Fall Apart,* placing this short story within a larger historical trajectory of Nigerian writing. Grace's work sorting through the musty archives is one that points to the archival turn in literary and critical theory. The archive contains a historical register, but it is up to the living historian to reimagine the past into life, to work through "the hard, obvious things that are printed in books and the soft, subtle things that lodge themselves into the soul."[43] In this way, Grace is a headstrong historian. So is her grandmother. The final image of Nwamgba's palm, "thickened from years making pottery," reveals a similar determination not to lose old customs and instead to pass on stories and practices to her granddaughter to ensure her own historicity is not lost in the face of global pressures of assimilation.

The different historians and historicities present in "The Headstrong Historian" point to a larger discussion of history's imbrication within questions of nature, class, race, gender, and power. Gayatri Chakravorty Spivak's famous question "Can the Subaltern Speak?" considers what voices contribute to historical narratives. Spivak details the lost histories of marginalized people, particularly women of color, that get excluded from official accounts of historical progress. Her work responds to the growth of postcolonial literary theory and resonates with the rise of globalization theory. Globalization comprises a vast complex of social, technological, economic, and geographic factors that threaten a new world order where history gives way to a hegemonic horizon. "Only by legitimating and valorizing ethnic and racial difference," writes Shaobo Xie, "can the dominated and marginalized peoples achieve recognition as equals and justify that equality, despite their insufficient degree of technological and infrastructural modernization."[44] Literary and media engagements with globalization can reassert a multitude of natural historicities in the face of global forces that consider history to be a fait accompli.

Such efforts to reclaim and resituate history find new relevance and urgency in an era of planetary change. Omar El Akkad notes the tendency in environmental thought toward privileging the future. As a result, memory, history, and the past may get lost in the scramble to confront the constant arrival of anthropogenic threats: "We must create new ways to think about what comes next, but also about what came before. As coastland drowns, as wildfires thin the ground and thicken the air, as changes that used to take centuries begin to take years, it will become increasingly difficult to anchor our memories to a geography, to a stable piece of land. So we must find other anchors—anchors that link memory to people, to relationships, to the solidarity and compassion and resistance that will serve as our only useful lifeboats in this storm."[45] Historians will answer this call, but Akkad places more importance on artists and storytellers for navigating the uncertainty that will grow between past and future.

Media and innovative digital projects will also contribute to this process of creating new anchors, lending new biases toward technologically advanced countries. Without attention to the reconfiguration of history, Akkad warns, a new radicalism will emerge in the wake of environmental dispossession. While extremism may be as undesirable as passivity and complacency, the energy of resistance, rebellion, and revolution will be necessary to ward off the more dire predictions for Earth's trajectory. The next two chapters consider the growth of environmental movements and media theory in the 1960s and 1970s in relation to the intersections of art, technology, and activism.

PART II

WHOLE EARTH RESISTANCE

3

From Global Village to Anthropocene

> Tomorrow we shall be able to look into the heart of our fellow-man, be everywhere and yet be alone.
>
> —László Moholy-Nagy

Environmental crises demand a dramatic return to reality. Wake up! Unplug! And look at what humans have done to the world. A new temporal sublime foregrounds the speed of changing planetary systems that have developed over millions of years. An instauration of technology, practices, and activism is necessary to make real change to renew and restore the biosphere. The divide between human and natural history collapses to reveal a proliferation of natural historicities. Chapter 2 explored how the Imperial Institute acted as an "empire story-land" to naturalize the subjugation of foreign lands and populations through an affective connection to products and people. While decolonization brought independence through the sweeping "winds of change," the promotion of global consumption has only increased in the late twentieth and early twenty-first centuries under the aegis of transnational corporations and neoliberal trade policies.

It is perhaps not surprising that a subject of the commonwealth first theorized a critique of imperialism and media. In his 1950 work *Empire and Communications,* the Canadian Harold Innis begins his inquiry into economic history by thinking about the trajectory of his homeland in relation to its natural resources, tracing the "succession of staple products that were demanded by the metropolis of empire . . . and the cultural interaction of the peripheral peoples that exploited them . . . fisheries were succeeded by the fur trade, timber, mining, wheat, oil, hydroelectric power, and so on."[1] When considering the role of timber in the pulp and paper trade, however, Innis connects the extraction of resources from the colonial periphery to the communications agenda emanating from the imperial center.

Just as the Imperial Institute moved from telling the story of products to telling the story of empire itself, Innis saw the use of mechanized media as a way to control space through the disruption of cultural continuity and history.[2] Innis's theories of communication proved more divisive than his more straightforward discussions of economic staples, but his work would profoundly influence his younger colleague Marshall McLuhan, who updated Innis's theories of papyrus and the printing press to consider the influences of television, radio, and the newly emerging computer.

McLuhan thought technology would create a "global village," a promising place for reasoned discourse, mutual understanding, worldly sentiment, and overarching agreement on the steps necessary to steward a better world. Yet technology has become no salve for the challenges facing species unification. Less a town hall than an echo chamber meets inquisitor's den, the discussions of environmental action facilitated by technology prove divisive at best and downright caustic at worse. Instead of a global village, the twenty-first century inherits a planetary *technosphere,* where all the weight and accumulation of consumption comes back to haunt local lives. At the same time, amid this material assault, art and creative intervention can become an "anti-environment," an *antimediator* that disrupts complacency and allows people to see their surroundings in a new light. If the techne of modern civilization has reached world-making potentials, where does the transformative power of art lie? What forms of poiesis can work to counteract the pollution and spoilage that humans have already brought forth in overabundance? Or is art just one more object in an increasingly unruly technosphere?

Noosphere Meets Technosphere

Alongside rising protests of chemical and fossil fuel industries led by Rachel Carson and others, artists sought new ways to engage with nature. In his extensive survey of land art, earthworks, and environmental installations of the 1960s and 1970s, James Nisbet identifies a through line of energy that animates ecological works. Energy as "a form of information" aligns these environmental projects with the concurrent proliferation of media networks, film, and the nascent cybernetic industry.[3] An ecological approach to aesthetics welcomes a comparative and additive understanding of forms—the more mediums the better for creating a robust dialogue. An artistic assemblage is needed to confront the forces of globalization. Yet Innis notes how empires also diversify mediums when at an impasse: "Introduction of a second medium tends to check the bias of the first and to create conditions suited to the growth of the empire."[4] The material accumulation of modernization combines with a holistic epistemology that seeks to comprehend the entire Earth system.

Around the time that the Imperial Institute was rebranding itself as an empire story-land, two French intellectuals, Pierre Teilhard de Chardin and Edouard le Roy (Henri Bergson's successor at the Collège de France), bandied about the term "noosphere" to describe a planetary layer that results from thought and reason. The term would later be popularized by Vladimir Vernadsky, a Russian biogeochemist who saw the noosphere as a final stage in nature's evolution, whereby humans as sentient beings enact nature's will. The noosphere, mindsphere, knowledgesphere, the know-us-sphere, but where exactly can we locate this sphere? Unlike the atmosphere, biosphere, or lithosphere, the noosphere has no physical presence. Instead, it becomes visible through its manipulation of these other existing spheres.

Although Teilhard coined "noosphere" in 1925, most of his writings were never published in his lifetime. His *Phenomenon of Man* was published posthumously in 1955, and its English translation appeared in 1959 with an introduction by Julian Huxley. As a founding member of the International Union for the Conservation of Nature as well as director-general of UNESCO, Huxley pursued many of Teilhard's ideas in relation to a global effort of environmental organization. Teilhard strikes a spectacular tone to imagine the noosphere as a layer of thought: "A glow ripples outward from the first spark of conscious reflection. The point of ignition grows larger. The fire spreads in ever widening circles till finally the whole planet is covered with incandescence. . . . It is really a new layer, the 'thinking layer,' which, since its germination at the end of the Tertiary period, has spread over and above the world of plants and animals. In other words, outside and above the biosphere there is the noosphere."[5] This passage resonates with the electrification of the planet, the glowing, interconnected grid of power observable from space. The spark of fire also nods to the era of emissions, the slow burning and release of carbon dioxide and other greenhouse gasses into the atmosphere, leading to global warming.

More than the image of the noosphere illuminating the top layer of the earth, however, the designations "over and above" and "outside and above" should give pause. Problematic eugenicist discourse employs this language of natural hierarchy. Teilhard and other proponents of the noosphere risk a similar reduction in cultural diversity, even while nominally affirming things such as the lack of a biological basis for race. The idea that global convergence will result in a human species that is interconnected, guided by ethical principles, and in agreement on how civilization should proceed remains a tantalizing goal. Noogenesis is alive and well under the guise of geoengineers who seek to shape and control the biosphere and atmosphere over, outside, and above. To read such aspirations after World War II must have seemed optimistic. To consider that Teilhard first formulated the noosphere after World War I seems even more aspirational. The *zone rouge* of World War I trenches puts the long-lasting effects of technogenesis in clear relief. Saturated with unexploded shells, ammunition, and heavy metals, the area was deemed uninhabitable and allowed to "return to nature," although the toxicity persists to this day.

Throughout his work, Teilhard equates genocide, industrialization, atomic fallout, and other undesirable consequences of development to mere growing pains for the more glorious dawn that will arise for humankind. "The future will decide what is the best name to describe the era we are entering," he writes with great prescience. "The word matters little. What does matter is that we should be told that, at the cost of what we are enduring, life is taking a step, and a decisive step, in us and in our environment."[6] That word, in contemporary times, is "Anthropocene." In one sense, it does matter little, but, in another sense, it matters greatly to stress the history, the technology, and the violence that define this ongoing process of becoming human. A former editor of the *Whole Earth Catalog*, discussed in the final section of this chapter, Kevin Kelly warns that the processes put in motion by noogenesis end up determining their own agendas:

> There is no electricity without the veins of coal mining, uranium mining, or damming of rivers, or even the mining of precious metals to make solar panels. There is no metabolism of factories without the ingest of food from domesticated plants and animals, and no circulation of goods without vehicles. This global-scaled network of systems, subsystems, machines, pipes, roads, wires, conveyor belts, automobiles, servers and routers, institutions, laws, calculators, sensors, works of art, archives, activators, collective memory, and power generators—this whole grand system of interrelated and interdependent pieces forms a very primitive

> organism-like system. Call it the technium. The technium is the sphere of visible technology and intangible organizations that form what we think of as modern culture. . . . Like all systems it has its own agenda.[7]

As the noosphere becomes weighted down by the rise of the material technosphere, humans must reckon with the autopoietic processes that have been set in motion beyond their control.

Beyond its material reality, the rise of the technium is a *hylomorphic* phenomenon, a compound of matter and form. Unlike the statue in Aristotle's analysis that arrives as a complete entity, today's planetary crises are ongoing and shifting around the poles of economics, political policy, and critical debate. This mediation is both a material process as well as a form of human expression of self-understanding as a species at different levels of representation, discourse, ethics, and social justice. Critics assert the difficulties of considering such a thing as a singular human species. By material measures, there are wide gulfs between levels of consumption for the lifestyles that have led to epochal change. For Aristotle, the measure of humanity as a species was the extent to which people can engage in rational, ethical discourse.[8] The legacy of Western philosophy's sociocentric discourse is taken up by William Connolly, who proposes an entangled view of humanism. Connolly questions how sociocentrism can be reconciled from its history of promoting the human species through the subjugation of nature. Instead of a hierarchical system, Connolly proposes a "pluralist assemblage organized by multiple minorities."[9] Rather than an assumed community, a disembodied noosphere, or a global village, Connolly considers an invitational "us" that would assemble stakeholders from around the globe to participate in an ongoing process of planetary mediation.

Art, literature, and media can become crucial reference points to help a pluralist assemblage navigate the maelstrom of the technium. Don DeLillo explores the legacy of De Chardin's noosphere in his short novel *Point Omega.* The book is framed by an extended meditation on Douglas Gordon's 1993 art installation *24 Hour Psycho,* which slows down Alfred Hitchcock's iconic horror movie to a run time of one full day. The central section of the novel follows an enterprising filmmaker, Jim Finley, who attempts to convince Richard Elster, a military consultant for the Iraq war, to create a confessional film in stark gray that would divulge a series of reflections and analyses of the war on terror.

Elster is a disciple of Teilhard's noosphere philosophy. He sees in the advance of consciousness a justification for expansion and self-determination, for war and violence if necessary. At the same time, war marks an omega point, an inversion where consciousness reaches an impasse and desires to return back to unthinking matter: "We're a crowd, a swarm. We think in groups, travel in armies. Armies carry the gene for self-destruction. One bomb is never enough. The blur of technology, this is where the oracles plot their wars. Because now comes the introversion. Father Teilhard knew this, the omega point. A leap out of our biology. Ask yourself this question. Do we have to be human forever? Consciousness is exhausted. Back now to inorganic matter. This is what we want. We want to be stones in a field."[10] Instead of the transcendence of the body into a realm of pure consciousness, Elster details an opposite movement, a retreat from the noosphere back into the realm of unthinking matter.

Elster's ruminations take place in the desert, itself an omega point away from the time swallowing cities. Within cities, however, art projects such as *24 Hour Psycho,* can present another type of transcendence through medial immersion. Here is an alternative omega point: "The less there was to see, the harder he looked, the more he saw. This was the *point.* To see what's here, finally to look and to know you're looking, to feel alive to what is happening in the smallest registers of motion."[11] At its best, aesthetic engagement can have this enlivening effect. No longer the opiate or mind killer, media immersion can raise consciousness to a level previously unobtainable. But to navigate this process, a larger discussion of technological extension and artistic intervention is necessary. A further transcendence is also on the horizon: a move deeper into technology itself that will accelerate in the age of internet computing, virtual reality, and artificial intelligence.

Extension and the Anti-Environment

Tom Wolfe points out how influential Teilhard's system of thought was on McLuhan's revolutionary media theories. Teilhard's noosphere became the "seamless web of experience," while the unified civilization of hominization became the global village.[12] Primitive man was both reduced through the anthropological gaze and romanticized for possessing "a more inclusive, totally sensuous commitment to situations."[13] Rather than a divine presence in technological progress, McLuhan substituted a shamanistic role for art and artists as navigators of the new environments created by medium and technology. Art, in short, could provide the right combination of spirituality and intellectualism: "The artist builds models of the new environments and new social lives that are the hidden potential of new technology."[14] Significantly, however, he positions art as a coping mechanism rather than a revolutionary channel for challenging new technological realities.

McLuhan cites William Blake as a paradigmatic artist who can illuminate the human environment in this fashion. Blake's portrait of Isaac Newton, for example, shows the scientist not in a lofty position, but, rather, crouched, bending over, sitting on a rock deep in a cave or on the ocean floor (see fig. 8). Critics often read this as a representation of Newton's alienation from the material world, but it might just as easily be read as placing Newton's body as a natural extension of the surrounding environment. His insights into gravity occur as he himself is weighed down, bent over, yet supported by the solid rock of matter. His left leg is obscured by the rock, leaving an amputated foot that seems to emerge from

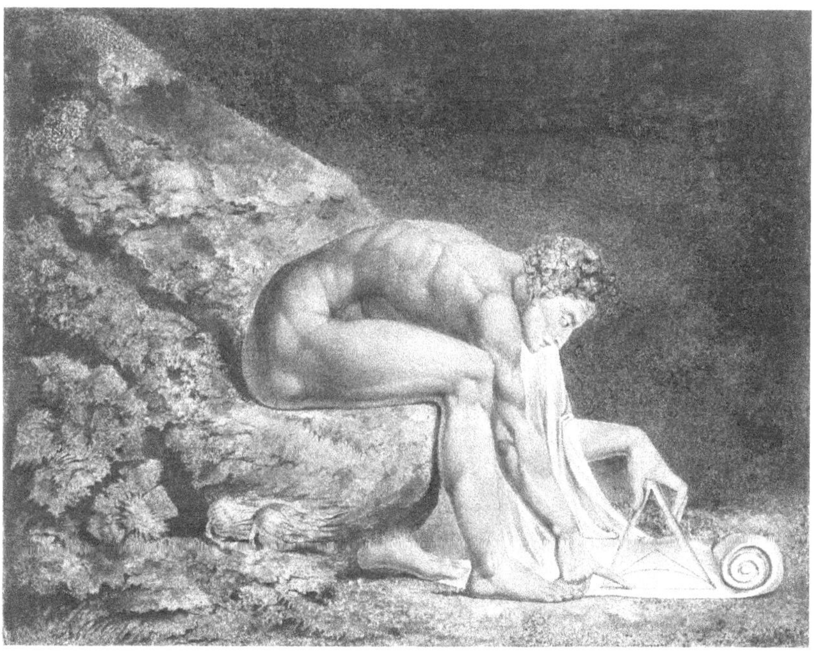

FIG. 8. *Newton,* William Blake, 1805. (Tate)

the stone itself. At the same time, Newton's intense focus does point to a certain escape from the primordial mire. McLuhan writes that "any medium, be it radio or be it wheel, tends to create a completely new human environment."[15] The designation of "human environment" is significant: technology creates new environmental possibilities, but not all creatures are equally invited to inhabit these spaces.

The "human environment" recalls Jakob von Uexküll's *Umwelt:* an organism's specific environment as determined by its biological receptors. Uexküll notes how humans are unique as a species in their ability to enlarge, expand, and qualify their existing biological receptors through the *merkzeuge* of technological innovations. A telescope, for example, expands vision dramatically.[16] He also warns, fatally, that no technology can allow a species to transcend biology. In his 1964 work *Understanding Media: The Extensions of Man,* McLuhan also considers the ambivalent nature of extension. Electricity, for example, extends the brain, but in doing so "suggests a desperate and suicidal autoamputation, as if the central nervous system could no longer depend on the physical organs to be protective buffers against the slings and arrows of outrageous mechanism."[17] There is a certain vertigo, a thrill to this out-of-body extension. At the same time, such an extended state might also cause panic, a bad trip. As a result, people tend to numb themselves to some degree to deal with the thrill of extension.

Exhilaration, anxiety, megalomania, apathy, myopia, detachment, alienation: such is the strange cocktail of psychological reactions to technological extension. While the focus is on electrical extension, the spatial order of urbanism can be understood in these terms as well. Anaïs Nin, for example, describes the phenomenon of extension in her story "Houseboat": "Now we were gliding along. I was running all over the houseboat, celebrating the strangest sensation I had ever known, this traveling along a river with all my possessions around me, my books, my diaries, my furniture, my pictures, my clothes in the closet. . . . A marine snail gliding through the familiar city. Only in a dream could I move so gently along with the small human heartbeat in rhythm with the tug tug heartbeat of the tugboat, and Paris unfolding, uncurling, in beautiful undulations."[18] This ecstasy is soon followed by mild panic as the houseboat starts taking on water, and the narrator must frantically begin bailing to protect her small world. To move in harmony with one's environment is both a thrill and an act of exposure. Only by reducing speed does the boat prove to be sound.

It is significant that electronic extension should move from the connected heart described by Nin to the dissociated brain. The neural network extends not so much through the sense of touch and rhythm, but by a more abstract, analytical process. Artists that wish to trace consciousness through this extension often make use of new techniques defined by the projection of light, a turn toward *energy* over matter. Extended far enough, with artistic prosthetics, McLuhan envisions a world of new architectural possibilities on an extrasensory plane:

> Painting with light is a kind of housing-without-walls. The same electric technology, extended to the job of providing global thermostatic controls, points to the obsolescence of housing as an extension of the heat-control mechanisms of the body. It is equally conceivable that the electric extension of the process of collective consciousness, in making consciousness-without-walls, might render language walls obsolescent. Languages are stuttering extension of our five senses, in varying ratios and wavelengths. An immediate simulation of consciousness would by-pass speech in a kind of massive extrasensory perception, just as global thermostats could by-pass those extensions of skin and body that we call houses. Such an extension of the process of consciousness by electric simulation may easily occur in the 1960s.[19]

Full extension requires not only artistic processes but also architectural design, atmospheric control, and eventually telepathic language. This passage presents the fully realized vision of a global village, one where humans have transcended their material bodies to arrive at an energetic harmony and understanding of one another. Some of these fantastic ideas are coming to pass. The global thermostat is the preferred metaphor for geoengineers who place faith in technology's ability to regulate planetary systems. While the world is heating, it is hardly in the controlled, dialed-in setting of a thermostat. Languages, too, are consolidating and simplifying in an increasingly globalized world. Yet, while communication may be easier than ever, largely through the dominance of global English, it is unclear if understanding and a shared consciousness is following suit. Often this increased communication only serves to polarize and divide people into contentious parties.

The Web of Malaise

It is important to keep in mind, as Glenn Willmot argues, that the global village was both paradise and inferno. Later in his career, McLuhan understood that for all the joys of technological extension, there were concurrent pains, anxieties, and terrors. Yet his faith in technology and his technodeterminism led him to undertheorize these downsides and stumbling blocks. Instead of the telescope that extends vision into outer space, Carson's *Silent Spring* highlights the "narrow window" of the electron microscope that reveals the molecular building blocks of life and the disruptions taking place there.[20] The narrowing window of scientific and technological disciplines also leads to the myopia of chemical industries that introduce a whole host of deadly consequences into ecosystems.

Against extension, one might place David Harvey's guiding metaphor of *compression* as the defining aspect of postwar global life. Harvey also stresses the increasing speed of modern life brought about by mechanical innovations and near-instantaneous communication, but instead of a lofty expansion, he characterizes this new state of experience as one where, "the world seems to collapse inward upon us."[21] One need not choose: instead, the planet may easily accommodate states of expansion and contraction, extension and compression; indeed, how else to explain the scope and scale of the ecological issues that continue to unfold? As an autopoietic phenomenon, the world expands and contracts, beating with the systole and diastole rhythm of consumption, expenditure, and belated regeneration. While energetic extension wraps around the globe, the material infrastructure required for this shining light leaves its ever-thickening imprint.

Some of the lightest and most ethereal technologies require the heaviest metals, the rarest minerals, and the most exhausting labor. Instead of the beautiful harmony of Nin's houseboat, what transpires in this autopoietic system is the "charge" of Plath's "Lady Lazarus," a toxic residue, a system of diminishing returns. Matter begins to seep back into the formal medium, the pure energy that seemed so transcendent does not thrill as easily anymore. Too much time spent on screens, in virtual realms, in mediated existence leads to dizziness, fatigue, nausea. Jean-Paul Sartre's iconic description of world weariness reveals the persistence of matter: "Objects should not *touch* because they are not alive. You use them, put them back in place, you live among them: they are useful, nothing more. But they touch me, it is unbearable. I am afraid of being in contact with them as though they were living beasts. . . . I recall better what I felt the other day at the seashore when I held the pebble. It was a sort of sweetish sickness. How unpleasant it was! It came from the stone, I'm sure of it, it passed from the stone to my hand. Yes, that's it, that's just it—a sort of nausea in the hands."[22] The nausea of Antoine Roquentin shows a world that is already materially extended. Objects are given agency in a way similar to more contemporary ecological theories such as object-oriented ontology and new materialism. Not only does matter extend out of itself toward the human species, but the individual is already extended into the environment. A firm sense of subjectivity is lost in the face of this overwhelming sensation.

Media such as film and television share in this spectrum of magical ecstasy and crushing ennui. Walker Percy's *The Moviegoer* updates Sartre's nausea in postwar America through a discussion of malaise. The novel tells the story of Jack "Binx" Bolling, a war veteran and erstwhile stock trader who seeks to avoid his ennui through a search for meaning in Louisiana. Binx recounts how he and some friends initially tried to work through PTSD with a good old-fashioned nature cure hiking on the Appalachian Trail. Unlike the post–World War I back to nature movement, however, this time the effort falls flat. Why? It is unclear, only that "the beauty of the smoky blue valleys, instead of giving us joy, became heartbreaking."[23]

Binx returns to Gentilly and, instead of nature, finds refuge in the movie theater. By contrast, his peak moment of malaise comes in an automobile: "On its way home the MG becomes infested with malaise. It is not unexpected, since Sunday afternoon is always the worst time for malaise. Thousands of cars are strung out along the Gulf Coast, whole families, and all with the same vacant headachy look. There is an exhaust fume in the air and the sun strikes the water with a malignant glint. A fine Sunday afternoon, though. A beautiful boulevard, ten thousand handsome cars, fifty thousand handsome, well-fed and kind-hearted people, and the malaise settles on us like a fall-out."[24] The passage portrays malaise as an atmosphere, seeping in through the windows. The attention to exhaust fumes and radioactive fallout speaks to less spectacular forms of pollution that nevertheless carry a toxicity all the more damaging due to its lack of visibility.

The ten thousand cars and happy, well-fed people are a result of the postwar American population and economic boom. What seems like pure prosperity, however, comes with ecological costs in the form of waste and consumption. The environment is left to soak up these remainders, just as the traumas of war continue to press around Binx's psyche. Binx details how his search for meaning moves from a vertical process of reading canonical books to a horizontal process of wandering the neighborhood and landscapes. Pursuing a sort of psychogeographic *dérive* (discussed in greater detail in the following chapter), Binx avoids malaise, but only for so long. While he can trace his rupture from the everyday to being wounded in the war, true rest evades his grasp. Films and movie stars provide a potential outlet, not necessarily of escape, but a glimpse of "an aura of heightened reality."[25]

More than other mediums, film, through its expanding industry, is particularly interested in world building. Whereas literature can describe an imaginary world, film, as Jennifer Fay discusses, pursues the actual construction of a world through the studio, the control of weather, and the façade of habitability. With this investment in "the creation of artificial worlds, unnatural and inclement weather, and deadly environments produced as much for the sake of entertainment as for scientific study and military strategy," Fay identifies the cinema as the "aesthetic practice of the Anthropocene."[26] If, as Amitav Ghosh argues, literature fails to accommodate the spectacular swings of climate change into its realistic narratives, film needs these dramatic shifts between its acts to heighten the drama and accelerate toward climax.

Artificial film worlds come to life through the moviegoing experience: unlike television which enters the home, the moviegoer must go outside, only to enter a dark room to experience a more complete immersion in the filmic world. Environmental consistency comes with a risk of depersonalization: if you cannot locate yourself based on your surroundings, if you might be anywhere, then it is only a matter of time before your sense of self and identity begins to slip as well. Percy's moviegoer recognizes this threat and actively works against it. While moviegoing itself is a journey into depersonalized spaces, Binx develops techniques to ground himself: he talks to the ticket-seller, to the owner of the theater, or his fellow audience members. It is the presence of other people, ultimately, that allows for self-differentiation. This simple fact is also a guiding principle of film: the good guy and the bad guy, the star-crossed lovers, and so on, but, because of its limited duration, film must deal in arcs and clichés. A fuller exploration of social dissolution occurs not in film but in television.

Telescoping Television

Environmental control in film functions at both the level of production and consumption, something that a television program cannot ensure. It is perhaps for this reason that the world building of television seems less important. Especially for the sitcom, the procedural, and other serialized shows, the backdrop is secondary. The stage set rarely changes, inviting comfort and familiarity with the televised world. McLuhan is interested in film's ability to transform architectural vision and spatial experience, but his greater interest lies in television: a cool, participatory medium that enters the domestic sphere and draws the viewer into its interconnected embrace. Jerzy Kosinski's 1970 novel *Being There* explores the allure of television. The novel tells the story of Chance, a gardener who improbably rises through the political sphere with banal aphorisms and a sort of everyman mutability. The opening of the novel details Chance's relationship to the television in striking fashion, worth quoting at length:

> Chance went inside and turned on the TV. The set created its own light, its own color, its own time. It did not follow the law of gravity that forever bent all plants downward. Everything on TV was tangled and mixed and yet smoothed out: night and day, big and small, tough and brittle, soft and rough, hot and cold, far and near. In this colored world of television, gardening was the white cane of a blind man.
>
> By changing the channel he could change himself. He could go through phases, as garden plants went through phases, but he could change as rapidly as he wished by twisting the dial backward and forward. In some cases he could spread out into the screen without stopping, just as on TV people spread out into the screen. By turning the dial, Chance could bring others inside his eyelids. Thus he came to believe that it was he, Chance, and no one else, who made himself be.
>
> The figure on the TV screen looked like his own reflection in a mirror. Though Chance could not read or write, he resembled the man on TV more than he differed from him. For example, their voices were alike.
>
> He sank into the screen. Like sunlight and fresh air and mild rain, the world from outside the garden entered Chance, and Chance, like a TV image, floated into the world, buoyed up by the force he did not see and could not name.[27]

The television in this passage is aligned with light rather than matter (this despite the incredibly bulky sets required to create the ethereal images). It defies the laws of nature, but Chance's experience as a gardener serves him in navigating this new world of images. As plants tend to the sun, so he has grown toward the light emitting from the cathode ray tube. The TV world is a Saturnalian one where up is down, light is dark, hot is cold. Yet, these juxtapositions are soft and harmonious: they invite Chance to mesh and merge with the images and people on the screen. Chance becomes a pliable plant in front of the television's benevolent rays. His entry into the mediated realm is portrayed as a new birth in naturalized terms. Nor does he leave the old world; instead, he forms a bridge between self and image, world and media, and this marks the smooth commingling that will accelerate his political and economic rise.

Chance succeeds with those he encounters out in the world not only because of his ability to speak familiar truths gleaned from his database of television programming, but also because he represents a malleable blankness onto which others can project themselves. Although he cannot read or write, he has gained an invaluable visual and social literacy through his constant consumption of television. Unlike the self-differentiation that Binx must labor to maintain between himself and others, Chance allows others to make of him what they will. He does not challenge their assumptions, and as a result they can endow him with qualities they find appealing.

Television is also distinguished from film in this novel through its ability to self-reflect and multiply in and of itself. Chance comes to this revelation when he is making a live TV appearance and visits a television control room: "Chance was astonished that television could portray itself; cameras watched themselves and, as they watched, they televised a program. This self-portrait was telecast on TV screens facing the stage and watched by the studio audience. Of all the manifold things there were in all the world—trees, grass, flowers, telephones, radios, elevators—only TV constantly held up a mirror to its own neither solid nor fluid face."[28] Here Chance experiences a moment of antimediation that allows him new insight into the medium he thought he knew so well. The machinations of television present a technological apparatus quite different from the black box of the film studio. Here television resembles more of a funhouse of mirrors that proliferates in narcissistic fashion.

Television falls in love with its own image. In Ovid's myth, Narcissus finds "substance / In what was only shadow."[29] While the scene is glossed as self-love and infatuation, it also bears significance that Narcissus rejects Echo, who is also another reflection of himself. While Narcissus is initially intrigued by the sonic repetition, he turns away when Echo reveals herself. When he sees her raised arms, he becomes adamant, like Sartre's Roquentin, that he should not be touched. Material contact proves to be the point at which he breaks away. At this spurning, Echo begins to lose her material form, wasting away until she exists only as auditory presence. Yet an echo still affirms the solidity of matter: it must bounce off *something.*

It is the shadow side, the darkness of narcissism, that Caravaggio captures in his late sixteenth-century masterpiece (see fig. 9). Like Jerome in the sepulcher, here Narcissus stares less into his own reflection than into the black void. This might be his image in an unilluminated television set. Caravaggio's chiaroscuro is the gold standard, the pure darkness that OLED screen technology attempts to recreate. Perhaps most striking in Caravaggio's work is the *knee,* a glowing protuberance, an orb in the void, a meteor flying through space, crumpled in on itself out of Narcissus's neglect. His absorption in the screen, in the void, comes at the expense of proprioception; he is no longer able to feel his own body. Out of obsession with self-image, Narcissus becomes detached from his own corporeality.

The face of television is like the surface reflection that beguiles Narcissus, a vision that is "only shadow, / Only reflection, lacking any substance."[30] Unlike the solid materiality of Echo's arms, it is some fourth type of matter: perhaps a plasma display that lends it magnetic qualities of enrapture. The result is an autopoietic loop that inputs reality and outputs virtuality. In such an environment, antimediation becomes more and more difficult. Both Binx and Chance engage with media to form a somewhat Faustian bargain with the ruling capitalistic order. Binx resigns himself to a life of stock trading and quiet comfort. Toward the end of the novel, he states, "Ten years ago I pursued beauty and gave no thought to money. I listened to the lovely tunes of Mahler and felt a sickness in my very soul. Now I pursue money and on the whole feel better."[31] Here the beauty of high art mediation leads only to nausea and malaise. For Chance, despite his meteoric rise through society and politics, one senses that he really longs to return to his life as a gardener, blissfully pruning plants and consuming the comforting rhythm of television.

How might mediation through light in TV and film lead to a transformational relationship with the ruling class order? A few years after

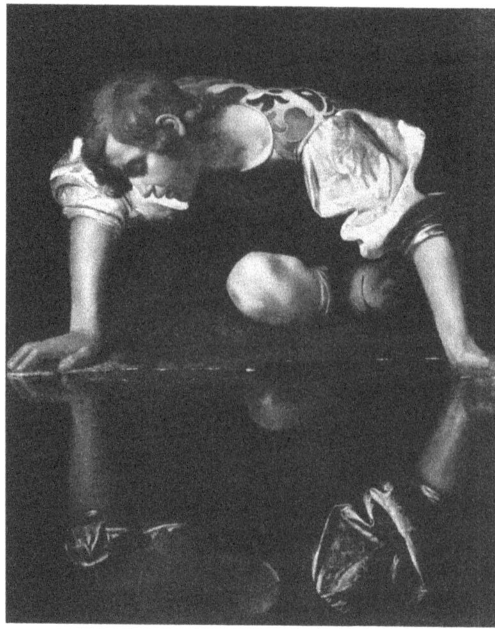

FIG. 9. *Narciso,* Caravaggio, 1594. (Gallerie Nazionali di Arte Antica, Roma, Bibliotheca Hertziana, Istituto Max Planck per la storia dell'arte/Enrico Fontolan)

Kosinski's novel, Nam June Paik released his iconic "Global Groove" as an "invasion" of broadcasting technologies. Paik began his artistic career as part of the Fluxus group of international artists. The Fluxus manifesto of 1963 declared a new confrontational mode that would "PROMOTE A REVOLUTIONARY FLOOD AND TIDE IN ART, Promote living art, anti-art, promote *NON ART REALITY* to be ~~fully~~ grasped by all peoples, not only critics, dilettantes and professionals."[32] The revisions and oxymoronic phrasing showcase how language itself strains to describe this new aesthetic reality.

Initially trained as a classical musician and composer, Paik's breakthrough came when he turned his sonic sensibility to the visual medium of television. His first solo exhibition in 1963 at Galerie Parnass, "Exposition of Music, Electronic Television," presented altered TV's and invited viewers to participate in manipulating the technology in new ways. By opening up the back of the television, Paik hijacked the "dictatorial medium" and encouraged people to "talk back to television" through the use of magnets to distort images from the cathode ray tube.[33] He made use of Sony's newly released "Portapak" handheld video recorder that promised to democratize video and proved highly influential for civil rights movements and antiwar protests in the 1960s. He also collaborated with Sony engineers to design a "Versatile Color TV Synthesizer" that would allow him to play the television like a piano. Describing Paik's work, John Cage writes that "he activates, timeifies, sculpture with video."[34]

Under the patronage of Howard Klein and the Rockefeller Foundation, Paik was able to bring his TV synthesizer into broadcasting studios to beam his new vision of art into homes throughout the United States. Promising a "landscape of tomorrow," Global Groove premiered on January 30, 1973 at 10:30 p.m. The project envisioned a media world where everyone would have their own channel and cultures might come together by tuning into any TV station all over the earth. His technological interventions liquefied the screen into new forms of malleability, blurring the lines between a real butterfly and a video butterfly. By turning things upside down on the airwaves of television, Paik offered a view of the world beyond the gazes of colonialism and corporate advertising. While aspects of his visual style would in time be coopted by these same interests to sell more products and produce more visually immersive forms of entertainment, Paik launched a wave of video art that continues to hijack and remix televisual images in novel and provocative ways.

Holistic Media for the Whole Earth

While many embraced the color picture box as the hearth in their suburban home, others sought alternative lifestyles away from filmic mediation. The environmental movement proliferated in part as a rejection of media pacification. The potential to reboot media and technology toward a more environmentally engaged future was pursued by the *Whole Earth Catalog* (WEC). First published in 1968, this modern-day Walden for the technological set pursued a self-reliant, ethical consumer culture for the denizens of Silicon Valley and beyond. Like the Imperial Institute, the WEC was omnivorous in its pursuit of materials, crafts, architecture, gardening, alternative energy, and many other subjects. The creators of WEC also invested in holistic aesthetics, combining print, images, illustration, and other techniques to create a unique media product that envisioned new ways of living in the world. While the Imperial Institute pursued a holism from above, seeking to unite the British Empire, increase global capitalist accumulation, and continue to enforce hierarchical positions, the WEC aimed at grassroots organization and anti-corporate alternatives.

Published by Stewart Brand, the WEC assembled an eclectic group of thinkers, makers, inventors, engineers, and artists to imagine an ecological future rooted in shared knowledge and group praxis. Drawing on Buckminster Fuller, Malcolm Wells, Norbert Wiener, and others to "imagine a synthesis between cybernetic theory and countercultural politics," the catalogue became something of a new-age georgic manifesto for cultural change that could rewire the brain.[35] The focus on the "Whole Earth" led to practices that could be deployed on small and local levels but also keep a global frame of reference in mind. More than anything, Brand was a curator of information, media, ideas, and goods. Alongside its investment in futurism, the WEC contained a primitivist strain with its interest in Native American tipis, Zen Buddhism, and tea ceremonies. There was also a tendency toward regressive, patriarchal norms within group projects. The 1971 film *Saline Valley Production* showcases men building geodesic domes while "women cook around the geothermal springs."[36] The WEC similarly gave short shrift to race relations and working-class life, although it did include conversations with James Baldwin and Margaret Mead, among others.

If the "Whole Earth" mission fell short in some areas, at its best it fostered a philosophical mindset that led to conscientious engagement with the world. The question remains, however, how ecological thought can avoid abstraction, and, in turn, how such abstracted thought might lead to misguided actions. A major part of the WEC aesthetic was envisioning the earth from afar. The cover was adorned with the planet viewed from space. Part of the draw for viewing the planet holistically from a distance was a new sense of *legibility*. Brand was drawn to photographers such as Hanns Reich, whose 1966 collection, *The World from Above*, presented "the history book of the world . . . huge and readable."[37] Where the Imperial Aviation efforts read resources and minerals from the skies, the WEC was interested in cultivating a sense of humility in the face of nature's sublimity.

Representation, however, is no guarantee for action or reparation. Nisbet presents the paradox succinctly: "The assumption, in other words, is that seeing the whole earth is tantamount to acknowledging the scale of late modernity's ecological crisis. But there is no such depth to pictures of the whole earth from space. Instead of broadening one's grasp of the earth's ecological predicament, they tend to simplify these systemic issues into an object, a flat disk. . . . What is missing from these visual representations of earth as an ecological object is a way to connect the experience of looking at an image of the entire earth with the sensory experiences of living upon it."[38] The famous *Earthrise* photo is an aesthetic encapsulation of the fragility of the planet, as well as the even smaller size of the human being, but it is also an abstracted vision of the world without humans.

To combat this abstraction, earthworks and land art can provide mediation between the experiencing subject and the vastness of territory. Nisbet cites Richard Serra's *Shift* as well as Michael Heizer's *Double Negative* as successful works precisely because they take up the question of the scale of human insertion into the earth. As such, they explode the traditional practice of sculpture to mediate "one's awareness of spatial position within the surrounding site."[39] Unliked the zoomed-out focus of space photography, here is a zoomed-in, somatic experience of landscape itself, one that compels an affective response.

While these large-scale earthworks lend themselves to readings of the sheer size of human influence, they are not necessarily the most effective communicators for reaching urban audiences. A work such as Heizer's *Dragged Mass,* on the other hand, wherein the artist used heavy machinery to tow a block of marble across the lawn of the Detroit Institute of Arts, brings the earth moving power of humans home to roost in the literal backyard of the art world (see fig. 10). The opposite of the pristine *Levitated Mass* that floats effortlessly in the air at LACMA, *Dragged Mass* reveals a petulant Atlas, no longer holding up the earth with stoic determination. The museum trustees had the block promptly removed and exploded at the end of the exhibit, with the lawn restored to its unmarred condition. However, the piece provided a drastic departure from isolated and unencumbering earthworks: presenting the viewer with a sense of weighty responsibility that goes along with world making technologies.

Rather than floating out into the ether, at its best the materialism of earthworks can direct viewers toward engagement with a land ethic. Excerpted in the pages of WEC, one of the leading practitioners of environmental holism was Aldo Leopold, who moved beyond aesthetics to pursue a holistic guide to practical activity. In 1949's canonic *Sand County Almanac,* he explains a simple land ethic that may serve to guide communities facing unexpected or sudden ecological changes. Leopold's land ethic "simply enlarges the boundaries of the community to include soils, waters, plants and animals, or collectively: the land."[40] Like much

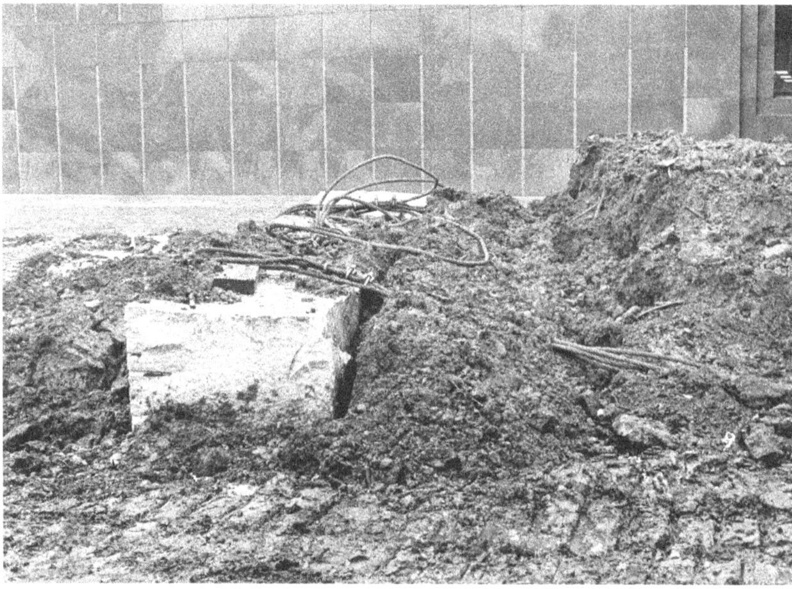

FIG. 10. *Dragged Mass,* Michael Heizer, 1971. 35 ton rock in removal of 300 tons of earth, cables, 5 feet 10 7/8 inches × 98 feet 5⅛ inches × 1 foot 11 5/8 inches (1.8 × 30 × .6 m). No longer extant. (Detroit Institute of Arts; © Michael Heizer, photo Michael Heizer studio, courtesy the artist and Gagosian)

holistic thinking, the expression of Leopold's land ethic is immediately persuasive and seemingly self-evident. On further inspection, however, the simplicity does not seem simple at all. And here the rub of holistic thinking is felt: too often it substitutes abstraction, intuition, or common sense in situations that are in fact more complicated. Like the transcendent truths experienced at the highs of a hallucinogenic trip that no longer seem so obvious in the sober light of day, holistic thinking is tantalizingly virtual. One need only process the land ethic through the much more tangled mediation of environmental justice for this to become clear. Land, in short, is never simply an autonomous entity: it almost always presents a complicated convergence of capital, property law, colonial violence, labor, and ecology.

One of the earliest commenters on the complicated nature of land ethics and environmental justice was Steven Byington, whose article "On Interference with the Environment" was serialized in the *New Freewoman* in 1913. Given the difficulties of protecting individual liberty from the sheer diversity of human interests, Byington suggests appealing to the environment as a mediator of right and wrong in human action. In the September 1913 issue of the *New Freewoman,* he writes, "If one person injures another by making the material environment unfit for that other's use, the injury should be regarded as on the same level with a direct assault on another's person or on the products of his labour. I say 'material environment', meaning such things as the air, the water, the hosts of birds and beetles and bacteria; not the social environment."[41] While this axiom seems simple enough in theory, in practice it quickly becomes complicated by the variety of competing interests present in any given environment. While the right to clean air and water seems straightforward enough, Byington admits difficulties when it comes to questions of personal property as well as certain animals that are less desirable than others. For example, he cites a bird, the bobolink, that is a blessing to one person's summer home and a curse to another's rice field. As for other animals such as deer, Byington considers it one person's liberty to hunt to the point of extinction just as it is another person's right to establish protected areas for wildlife.

What tempers a holistic aesthetic is the ethics of practice. And, to the credit of the WEC, application is continually on display. A number of articles and book excerpts reprinted in the catalog pages consider how the individual can navigate the demands of mind and body to interact with culture and environment. Edward T. Hall's *The Hidden Dimension* outlines a program of proxemics to raise the sense of touch above sight and hearing. Focusing on proprioception, the kinesthetic feedback function, Hall explores how sense of self is wrapped up with proximity to other people, temperature, and a whole range of feedback mechanisms.[42] Similarly, Norbert Wiener's *Cybernetics* provides an informatic basis for the engagement with environment:

> Man is immersed in a world which he perceives through his sense organs. Information that he receives is co-ordinated through his brain and nervous system until, after the proper process of storage, collation, and selection, it emerges through effector organs, generally his muscles. . . . Information is a name for the content of what is exchanged with the outer world as we adjust to it, and make our adjustment felt upon it. . . . The needs and the complexity of modern life make greater demands on this process of information than ever before, and our press,

modernists left him too focused on form over matter. He celebrated a hollowed-out medium, detached from labor and the natural resources that have become increasingly sapped through the global extension of telecommunications infrastructure. Across the Atlantic, however, a group of French intellectuals, artists, architects, and activists were tackling these questions to understand how social revolution might disrupt the ruling capitalist order. The next chapter will explore the possibility for ecological rebellion by rerouting or hijacking the status quo in the Situationist practice of détournement.

4

From Ecological *Détournement* to Rebel Geology

> I rebel—therefore we exist.
>
> —Albert Camus, *The Rebel*

ACKNOWLEDGING THE scale of human impact on the planet is the first step toward broaching the much more difficult question: What is to be done? Anointing the geological era of humans invites further intervention. A positive response is to look at current practices, to consider what can be improved and made more efficient. Toxins can be identified and minimized. Habitat restoration in the twentieth and twenty-first centuries has become a robust practice that has seen major transformations for land throughout the world. New laws can protect species on the brink of extinction and bring populations back to health. All of these environmental agendas are laudable and reflect many years of involvement at local and international levels in community organizing and political advocacy. Yet the pace of changes, the acceleration, leads to the question of whether these moves toward conservation, sustainability, renewal, and restoration are enough.

Corporations can increasingly don the veneer of environmental stewardship through greenwashing, carbon offsets, and publicity campaigns. Thirty thousand feet above the earth, one can smile at a reassuring message on a napkin that air travel is now "carbon neutral" before blithely crumpling it up and throwing it into the trash bag of a passing flight attendant. To stem the tide of the rising technosphere, a larger, more *revolutionary* movement may be necessary. Yet how is this to proceed? Revolution as political upheaval requires a specific target: a monarch, a dictator, an imperial power. The need to rebel, to stand apart from the status quo is more imperative than ever to counter the increasing passivity and acceptance of the ruling order encouraged through media consumption. At the very least, protest brings people face-to-face with one another. At a certain point, one must turn off the television, close the laptop, and walk out into the streets.

Literature, art, and media can contribute to change, can connect people in discussion and create urgency around important issues. But these forms can also mediate social agitation back to an acceptance of small concessions and minor gains. Environmental themes and messages have become commonplace in novels, movies, artworks, and video games, but the channels through which people view ecological crisis are not necessarily the best ones in which to fight back. This chapter looks at the critique of media and technology led by the Situationist International as well as the avant-garde collective Utopie surrounding the protests of 1968—a widespread movement that galvanized people across multiple continents. Instead of violent upheaval, a sustained engagement with environmental issues forms the basis of a rebel geology, one that draws on a long history of political engagement to chart an equally long future for biodiversity.

Are You Mediated?

There has been a revival of interest in the Situationists and the practice of psychogeography in relation to contemporary environmental issues, led by the London Psychogeophysics Summit and the publication of their manifesto in *Mute* magazine in 2010. As a rebuke to the urban focus of psychogeography, the geophysical foregrounds instead how the "human condition is being shaped by the entire earth."[1] While this recognition of human finitude and frailty in contrast to the nonhuman may be convincing from an aesthetic point of view, the political stakes of such a program are conspicuously lacking. Instead of action, the London group calls for awe and terror in the face of the larger-than-life forces of plate tectonics and magnetic fields. The manifesto ends with a decidedly retrograde championing of Joseph Conrad's aesthetics of fear and wonder at the strangeness of the Congo jungle in *Heart of Darkness.*

While innovative in some regards, this psychogeophysical preoccupation enjoys a longer history as well, stemming back to mesmerism and theories of animal magnetism. One of the earliest works of climate fiction, Jules Verne's *The Purchase of the North Pole* or *Topsy Turvy,* published in 1889, hinges on the inability of humans to reorient the axis of Earth's rotation. This story presents the scheme of the Barbicane Corporation to purchase territory in the Arctic and then use a massive gun to realign the earth's axis to be horizontal, thereby melting the glaciers and exposing landmasses for coal mining.

A new, horizontal axis will usher in an epoch of the "earthly globe," in which temperatures will be moderate and advantageous to all sorts of human activities for enrichment. The company enacts the scheme by boring a cannon into the cliff of Mount Kilimanjaro, but the explosion proves ineffective at altering the tilt of the earth's axis. A mistake of calculation is cited as the culprit, and the real force would require a trillion such cannons. The story concludes with a note of reassurance: "The inhabitants of the earth may sleep in peace. To modify the conditions in which the earth is moving is beyond the efforts of humanity."[2] This basic argument is the premise of much climate change denial, and it is one that the Psychogeophysical Society repeats, if perhaps unintentionally. At the same time, the society is on to something in their desire to revisit and reevaluate Situationist practices in relation to anthropogenic changes.

The work of the Situationists emerged at a crucial juncture between urbanism and media, lending their theories of alienation and passivity lasting relevance for the digitally saturated present. It was serendipitous that new technologies of television, the personal computer, and the mobile smartphone arrived in quick succession to save the modern subject from the boredom of the perfectly controlled, hermetically sealed environment. The open, white-walled boxes of modern architecture find their ideal antidote in the black squares and rectangles of screens. The saturation of media in contemporary times separates individuals, creates a false sense of connection, and discourages community activism. Beyond urban environmental aesthetics, it is this interpersonal component that remains important for thinking about mediation.

While the SI may have exaggerated their involvement in the 1968 rebellion, their environmental theories were nevertheless tied to social activity. While McLuhan pursued a technodeterminist worldview where enlightened man might embrace his destiny, Guy Debord and the Situationists brought a Marxist analysis of materiality and class relations to analyze media's role in promoting apathy and acceptance of capitalist ruling order. Debord seemed to both respect McLuhan's technical analysis of media systems and at the same time scorn his steadfast agnosticism in the face of social issues.

Debord's wariness was inherited by Jean Baudrillard, who developed a more robust critique of media's social influence and its limits as revolutionary technology. Baudrillard began his theorizations in the pages of

Utopie and would eventually embrace a postmodern articulation of simulation and virtuality to, in the words of Andrew Hussey, grapple with an "external paralysing force which is truly international in its ability to melt frontiers away and reduce all particular individual experience to phantasmic reality."[3] As early as 1970, Baudrillard praised McLuhan's diagnosis of the formal elements of media while condemning his "delirious tribal optimism."[4] It was in the aftermath of the May 1968 rebellion that Baudrillard was able to combine a critique of media with prevailing theories of space and urbanism, resulting in a uniquely environmental media theory that would continue to define his cultural criticism of the information age. To understand how he got there, however, one must first turn to Debord's *The Society of the Spectacle* and the revolutionary praxis of the SI.

Instead of extension, Debord's guiding metaphor was separation and enclosure. Like the medieval enclosure of the commons, the media spectacle divides and conquers to constrain consciousness, reducing it to a one-size-fits-all psychological response. Against this threat, the Situationists pursued psychogeography, the *dérive,* and détournement: the reconfiguration of existing space, time, art objects, and texts. Early on in 1967's *La Societé du Spectacle,* Debord clarifies that "the spectacle is not a collection of images; rather, it is a social relationship between people that is mediated by images."[5] This description is paradigmatic of Debord's theoretical formulation for the spectacular society he will go on to elaborate through 221 strophes that range from maxims to historical diegesis to materialist critiques of labor and commodities.

Debord's early invocation of mediation gives a clue into how this topsy-turvy world might function. The passive construction of the phrase is also telling. The French, *médiatisé,* belies a strange docility that allows interaction to occur. To be mediated is to connect via an intermediate matter in order that one thing may transmit to another. One can mediate as a form of active intercession, but to *be mediated* implies a subtle acceptance to a channel. For McLuhan, media allowed "man" to extend himself, in typical masculine colonialist fashion. For Debord, however, media just as often constricts, compresses, and conforms the viewer. One point of shared interest, however, lay in the automobile, a machine unique in its ability to transform individual mobility as well as social space. Speaking of the disaster that is metropolitan sprawl, McLuhan concedes, "a technological extension of our bodies designed to alleviate physical stress can bring on psychic stress that may be much worse."[6] Debord, for his part, calls the automobile the "supreme good of an alienated life and essential product of the capitalist market."[7]

The automobile has been theorized and aestheticized perhaps more than any other machine in the twentieth century. Few have arrived at its essence in so succinct a fashion as the great critic of Conrad's surreptitious prose, Chinua Achebe. His 1971 poem "Benin Road" exemplifies the dualities and divisions of separation through the juxtaposition of nature and technology. A paved road cuts through the forest where a butterfly floats effortlessly. A car speeds along, bound by the asphalt conduit. What happens next is a dramatic encounter:

> I come power-packed enough
> for two and the gentle
> butterfly pops open
> in a bright yellow
> smear in the silicon
> hardness of my vision[8]

Here the driver is indeed extended. But the silicon windshield not only expands the driver's vision but in fact becomes a medium through which to view the world differently. As for the butterfly, it experiences an unfortunate moment of détournement, an explosion where inside becomes outside. What was once demure and weightless becomes vivid and brusque. On one level the poem is a simple allegory for technological progress. Modernity arrives with a force that brushes nature to the side. But the collision also reveals a new aesthetic order: the bright yellow smear is like an abstract painting, an unexpected brushstroke that delights and arrests. The poem is also, more pointedly, an allegory for the Nigerian Civil War. Here the smeared butterfly represents the bright yellow half-sun on the Biafran flag. Now the violence is not simply modernity but also history—a recurring and rippling effect that whiplashes the process of decolonization into eruption.

Achebe includes "Benin Road" in his 2004 volume of *Collected Poems;* however, he presents a revised version. Now the butterfly undulates rather than fluctuates. The "paved forest tunnels" are more explicitly labeled a "brash new highway." But perhaps the most interesting change, from the perspective of technological mediation, comes with the revision of the final lines, where the "gentle butterfly offers / Itself in bright yellow sacrifice / Upon my hard silicon shield."[9] What was once a vision—hard but necessary—has become a shield, a defensive posture against the outside world. What was once extension has become retreat. The butterfly sacrifices itself, a martyr instead of an innocent bystander. Time has mellowed the poem a bit. Ever one to draw on classical mythology, perhaps the shield here gestures at W. H. Auden's "The Shield of Achilles," which portrays "an artificial wilderness / And a sky like lead."[10]

The promise of technological extension begins to ring hollow in the twenty-first century. Even as technology becomes more like the butterfly, ephemeral, light, and nimble, a calcification occurs. What starts as a vision of the future eventually becomes a shield that guards the past. The Situationists straddled this tipping point through the protests of 1968 and into the early 1970s, bringing new techniques of resistance to bear on a world increasingly programmed, sheltered, and isolated under the great compromise of the spectacle.

Mondialisation and the Geoengineer

The urban aesthetic theories of Debord and the Situationist International provide a significant vantage for coming to terms with human-dominated ecosystems. Anthropogenic climate change presents the atmospheric and geologic realization of what Debord terms the "second nature" that imposes "inescapable laws upon our environment."[11] While not expressed in geologic or ecological terms, Debord's *The Society of the Spectacle* invokes a Marxist historicism wherein natural history cedes to human history: "'History itself,' says Marx, 'is a *real* part of *natural history,* and of nature's becoming man.' Conversely, the 'natural history' in question exists effectively only through the process of a human history, through the development of the only agency capable of discovering this historical whole; one is reminded of a telescope, whose range enables it to track the retreat of nebulae *in time* toward the edge of the universe."[12] One of the defining features of the Mediacene is precisely the self-reflective understanding of human influence on natural history through the use of technology and the sciences.

It is at this moment that a spectacular valence enters humanity's understanding of its own position within the world: human history not only reveals natural history but begins to shape it as well. At the same time, humans are not necessarily in control of this process; rather, human activity takes on a life of its own. As Debord writes, "The society that molds all of its surroundings has developed a special technique for shaping its very territory. . . . Urbanism is capitalism's seizure of the natural and human environment; developing logically into absolute domination, capitalism can and must now remake the totality of space into its own setting."[13] This spatial seizure does not distinguish between the natural and the built environment. As such, people become subject to an environmental order of their own creation that nevertheless does not always act in the best interest or at the will of its members.

If the Situationists and Debord ultimately reached an impasse when applying their environmental theories to urbanism, their work nevertheless remains well suited to discussing lived experience and sensation during the contemporary moment of global climate change. Theories of détournement, in particular, what Greil Marcus calls "a radical deconditioning: to demystify their environment and the expectations . . . brought to it," seem more pressing than ever if the current carbon emissions trajectory is to be changed.[14] McKenzie Wark outlines détournement as a way of liberating the past and history as a new commons, free for the taking and remaking.[15] Yet such détournement can be difficult to achieve as capitalists invest more than ever in restricting the commons and maintaining business as usual. Jason Moore proposes the term "Capitalocene" in order to position "capitalism as *world ecology*," where nature as both material resources and human labor supply is notoriously devalued.[16] Disrupting this mentality of "Cheap Nature" becomes the chief goal for achieving a "different ontology of nature, humanity, and justice."[17] While Moore believes the material limitations of the planet will ultimately put a stop to cheap nature capitalism, Naomi Klein's *This Changes Everything* has more confidence in the adaptability of exploitation. She notes the close ties between capitalism and certain brands of environmentalism: As awareness of climate change grew in scientific communities, those in the extraction business have invested more resources into shaping the narrative.

With its move toward neoliberal globalization, capitalism becomes increasingly able to profit from environmental management. While Klein's argument is formidable on such a macro level, it becomes a bit more tangible when she describes effects on an individual and social level: "Contemporary capitalism has not just accelerated the behaviors that are changing the climate. This economic model has changed a great many of us as individuals, accelerated and uprooted and dematerialized us as surely as it has finance capital, leaving us at once everywhere and nowhere. These are the hand-wringing clichés of our time—What is Twitter doing to my attention span? What are screens doing to our relationships?—but the preoccupations have particular relevance to the way we relate to the climate challenge."[18] Klein does not fully explore this line of inquiry, but the passage notably shares many similarities to a Debordian society dominated by spectacle. While the proliferation of social media, wireless connectivity, and smart devices will take center stage in the last two chapters of this book, of chief interest here for both Debord and Klein is the rise in apathy, alienation, and complacency that accompanies everyday life in the urban environment.

The built environment becomes the integrated machine of modern capitalism and the playground of modern art. Published a few years after Debord's *The Society of the Spectacle,* Manfredo Tafuri's *Architecture and Utopia* explores the tensions and unexpected complicities between capitalist development and artistic expression. Writing after the protests of 1968, Tafuri is sensitive to the ways in which urban space can catalyze or stymie social action. He cites M. A. Laugier, who saw the development of the city as a natural phenomenon guided by the aesthetic of the picturesque.[19] Laugier proposes the model of a city like a forest, guided by variety, novelty, and surprise, while maintaining a harmonious whole. In practice, however, architectural naturalism, as Tafuri points out, was able to sidestep the inorganic and artificial challenges of urban development by neglecting social issues under the banner of natural progress. The close connection that this urban naturalism shares to picturesque landscape aesthetics leads Tafuri to see an ideological role in artistic activity that creates "new sublimations, rendered artificially objective by means of the call to the universality of Nature."[20] In Tafuri's analysis, art and literature are not immune to the larger environmental strictures of capitalist society; instead, they become the handmaidens that help the enlightened subject process and accept new environmental realities. The Situationists—whom Tim McDonough identifies as falling under the banner of Tafuri's critique of the avant-garde as instrumental to the restless revolution of capitalist society—wrestled nevertheless with a further iteration of naturalism as ideological control that emerged as the built environment began extending into the representational realm of spectacle.[21]

One can detect this same Enlightenment naturalism creeping into discourses on the Anthropocene, where the expansion of human influence can be framed as a natural development of the same forces that built and connected the great cities of global modernity. Such a logic reaches its apex in calls for humans to embrace their new role as epoch makers through geoengineering, the dream of technological fixes that aim to create a global thermostat. Art, literature, and media can change humanity's attitude toward the climate crisis, but by Tafuri's logic such efforts might equally be a way of assuaging anxiety and fear of the new world order. This is a hallmark of mediated life: there are unprecedented ways to visualize and aestheticize ecological crisis, yet such activity draws us even deeper into capitalism's spectacle of anthropogenic power.

It is perhaps only the monstrous return of nature itself—changed, augmented, made weird by human activity—that can disrupt and *détourne* the human-dominated ecosystem. It takes the arrival of a Hurricane Sandy or historic droughts that thrust society back into reality to discuss changes to urban planning and emissions futures, even if such efforts are incomplete and short lived. Rather than waiting for disaster to arrive, many position the arts as a training ground for confronting environmental issues in a proactive manner. Ian Baucom sees a compelling case for new modes of environmental justice stemming from literary encounters. For Baucom, fictive characters—especially protagonists of historical novels—give readers the capacity to move beyond a narrow sense of self and realize a fuller understanding of their position in the world. Baucom poses the aesthetic as ultimately a question of being. "'Experiencing' the nondisjunctive plurality of human life across those multiple forms of existence," he writes, allows one "access to a new conception of justice for, within, and against the looming 'inevitability' of the Anthropocene future."[22] Here, again, the city seems like a logical starting point for investigating the collective constitution of modern plurality.

Tafuri identifies Paris as the likely model for Laugier's city as forest. The following section explores life in this urban milieu as portrayed in Patrick Modiano's *In the Café of Lost Youth*. This short work engages the emerging aesthetics of Situationism while locating it within the restless and alienating realm of capitalist modernity. The fictive remediation of the SI movement provides an opportunity to question its theories and social commitments. The protagonist, the elusive Louki, is precisely the type of character that Baucom sees as full of potential. Modiano structures the work around denying the reader ultimate access or "truth" to this character. His inclusion of four narrative perspectives presents a plurality that undercuts and antimediates the authoritative voice of the historical novel. Along the way, a new environmental aesthetics emerges that emphasizes the importance of affective connections between humans and nonhuman elements of the environment.

Fixed Points and Neutral Zones

Modiano's *In the Café of Lost Youth* explores the connection between psychic geography and the urban environment. The setting of the novel is unmistakably Paris, but Paris is less on display than a certain urban experience. For the titular youths, there is little that seems to define their identity other than location. They are, quite simply, an assemblage of individuals who meet at a specific site (the café Condé), modeled on a café in the Latin Quarter frequented by Debord and other members of the Situationist International. They are regulars: it is repetition that defines their community. Yet this is not to say their existence is random. The café itself is not treated as what Marc Augé terms a non-place. Indeed, the Condé is so specific, so particular, that the first narrator claims one could easily identify its neighborhood and social milieu even if entering with a blindfold.

The café, then, is far from inhabiting a "neutral zone," as defined by the final narrator Roland. It is closer to a "fixed point," what Bowing, "the Captain," terms a refuge "amidst the maelstrom that is a large city."[23] Yet this is not quite it either, as the fixed points are not so much places but rather moments at a specific location and time. Bowing keeps a notebook wherein he records the customers who come and go in the Condé. This record is contrasted to photographs, which also locate the lost youths years later in a monograph published on Paris.[24] If the photographer, himself a Condé regular, fixes the scene through images, Bowing pursues a psychological role, tapping into the memory and imagination of social relations. While the photograph or the list of fixed points might, in theory, endure perpetually, a leather shop eventually succeeds the Condé itself.

As much as Bowing's fixed points appear as resistance to the faceless, alienating march of urbanization, they also resonate with other, unsettling hallmarks of modernity: surveillance, security, and data. An elder patron of the Condé, Dr. Vala, notes as much when he compares Bowing's notebook to a police register. Critics have noted the strange sympathy between the Situationists and the modern police state. As McDonough points out, "What was . . . proposed in common by the Situationists and 'the police' was nothing less than complete integration of the subject with the mechanisms of socioeconomic rationalization, with what the SI called, in proper Marxist phraseology, 'the domination of nature.'"[25] The fixed points, in particular, suggest contemporary technocapitalism that operates through smart phones, GPS, location sharing, tags, and check-ins to fix the consumer spatially. These are the user-based technologies that extend the project of nature's domination to human populations. Holding an interface to the world in the palm of one's hand achieves, in Tafuri's words, "the voluntary and docile submission to those structures of domination as the promised land of universal planning."[26] Against such a fate, the SI proposed a new form of psychogeography that called for the "destruction of the subject" in favor of a vision of a "new, malleable humanity."[27] Louki takes such a program to an unfortunate extreme through her suicide. Renouncing her subjectivity, she throws herself out of a window onto the (decidedly inflexible) streets below.

The regulatory aspect of the fixed point develops through the second section of *In the Café of Lost Youth,* where the private investigator Caisley manages to locate Louki, only to leave her in freedom. Significantly, it is an actual police register—and the ensuing work of police officer Bernolle—that leads Caisley to Louki's home address in Paris. Louki, whose legal name is Jacqueline, earns her place in the police registry for the minor infraction of juvenile vagrancy, a particularly telling offense in relation to fixity. Caisley imagines the crime in strongly environmental terms: "Juvenile Vagrancy . . . those two words evoked for me a meadow beneath the moon, beyond the Caulaincourt bridge all the way back there, behind the cemetery, a meadow where at last you could breathe in the fresh air."[28] The imagery here—similar to Louki's own narration yet also slightly different—is notably pastoral in nature. It is geographically accurate, describing Caulaincourt and the Montmartre cemetery, but also imaginative, describing an open meadow that seems to exist only in Caisley's own psychic geography. The open meadow provides the antidote to the urban maelstrom bemoaned by Bowing and the regulars at the Condé. Caisley feels a strong sense of empathy toward the young Jacqueline, such that he abandons the employ of her husband Jean-Pierre Choureau in favor of maintaining Jacqueline's autonomy as the newly renamed Louki.

This alliance between Caisley and Louki, one sided and unknown, comes to replace the institutional sanction of marriage between Jacqueline and Choureau. At the close of his interview with Choureau, Caisley reflects on such alliances, reformulating the logic of Bowing's fixed points as significant precisely due to their arbitrary nature: "In this life that sometimes seems to be a vast, ill-defined landscape without signposts, amid all of the vanishing lines and the lost horizons, we hope to find reference points, to draw up some sort of land registry so as to shake the impression that we are navigating by chance."[29] The remaining two sections of the book—the perspectives of Louki and her lover Roland—showcase a search for freedom and meaning within the vague urban landscape, both physical and psychological. Louki's narrative largely proves Caisley's conjecture that she desires escape, to "put herself out of reach for good."[30] Yet her most significant moment of escape comes not in some secluded, out-of-the-way meadow; instead, she achieves relief from climbing to the top of Montmartre. In the process, she compares herself to those (men) who climb the Tibetan mountains in search of Shangri-La: "For me, Montmartre was Tibet . . . Up there . . . I could truly breathe for the first time in my life."[31] In this narration, Louki's escapism takes on a quality different from the secluded meadow that Caisley imagines. The isolated mountaintop symbolizes power and dominance rather than weakness and fragility. At the same time, the escape, the breath of fresh air, is described in equally environmental terms.

Louki takes on the role of the romantic wanderer, climbing to the top of the mountain in search of the sublime, for a moment of transcendence that is the hallmark of a disinterested environmental aesthetics. As if to reinforce this point, Louki compares the intoxication she feels atop Montmartre to the feeling of breaking ties with another person. Yet these moments of escape are not enough for Louki, who desires a further environmental transcendence with nothing but "blue sky and the void."[32] Louki tests the extreme logic of disinterestedness: letting oneself go to the extent of leaving the world entirely. Roland also shares an interest in the void; instead of emptiness, however, Roland explores "dark matter," substance that is invisible and undetectable by humans yet nevertheless exists. His interest in neutral zones stems from an anthropological expression of this fundamental element of physics: a hidden manifestation of the universe's secrets.

If the fixed points work through identifying people in a specific time and place, the neutral zones encompass generality: a region, a set of out of the way streets, a nondescript building divided into furnished apartments that require little in the way of a tenant's history or background. The neutral zone is the opposite of the fixed point, yet Roland's method for outlining said zones is noticeably similar to Bowing's register of fixed points: the incomplete and ultimately aborted text *On Neutral Zones* amounts to little more than a list of streets and neighborhoods where said zones purportedly exist. Both Roland and Bowing imbue the neutral zone and the fixed point with certain talismanic properties of protection: the fixed point represents a life vest in the maelstrom of modernity, the neutral zone a protected cove safe from the storm.

Roland's own desire for fixity becomes clear in the most ecologically charged scene in the book. Reflecting on the past in the present, several years after Louki's suicide, Roland has a chance encounter with the mystic Guy de Vere.[33] The Condé café is now the Prince de Condé leather shop. Roland adopts a carefree attitude, a performance of the summertime romanticism he once enjoyed with Louki. His attitude shifts, however, when he encounters a tree: "A notice was attached to the trunk of one of the great trees whose leaves offer shelter on the way to the entrance to the gardens further along at Saint-Michel. 'DANGER. This tree will be cut down soon. It will be replaced by a new one this coming winter.' For a brief second, I thought I was having a bad dream. I stood there, frozen, reading and rereading that death warrant . . . In this world where I felt more and more like a holdover, the trees were on death row, too."[34] In a strange sense, this passage is the emotional center of this elliptical work, a profound moment of détournement that forces Roland to reevaluate his entire environment. Roland's connection to this tree is detailed with more pathos than any of the other narration spent on the lost youths. This moment carries with it the weight of Louki's suicide—the loss of a life that cannot be replaced—as well as the generation of lost youths who passed through the Condé.

The condemned tree encapsulates the domination of nature pursued by enlightenment modernity, the industrial revolution, and capitalist urbanization. Further, the fact that the tree will be replaced, points toward contemporary promises of geoengineering a sustainable future. Louki and Roland seek to escape the overbearing aspects of the human-dominated environment. Louki does this first by abandoning her husband, whose occupation, fittingly, is partner in the Zannetacci Real Estate Company. The husband Choureau represents the capitalist system of private property that begins with the division between those who own the land from those who pay to inhabit it. His own ground-floor apartment in Neuilly, where Louki abandons their marriage, is the embodiment of a lifeless, suburbanized architecture.

If this faceless system of private property—ownership of the earth—provides one pole, the École des Mines, an institution sheepishly attended by the novel's unnamed first narrator, provides a more utopian model of advancement.[35] An embodiment of Francis Bacon's *New Atlantis,* the École des Mines provides the ideological and technological foundations for mankind's expanding domination of nature. Those behind the tree's removal and replacement are likely graduates of such an institution. The novel's first narrator hides his enrollment in the École des Mines from his friends at the Condé, who prefer a romanticized education in the streets of life. By the end of the book, however, Roland recounts how he and Louki convince the young student to abandon his studies.

Economics and Science are the twin poles of anthropogenic power that shape the environment that is Paris, France, and the larger Western-dominated global ecosystem. What escape is possible from such a world? Does the student's truancy signify a small moment of rebellion? Lost youth, in this sense, signals not just the passing of adolescence, but the passing of alternatives to an increasingly oppressive social order. Does the suicide that is both the presence and absence at the heart of the novel stand for a larger renunciation in the face of implacable economic forces? Is there no tonic or effective remedy to environmental unease?

Such questions, of course, have no answer, and are precisely the stuff that drives the narrative uncertainty of Modiano's text. One is encouraged to cross-reference the narratives against one another, to find points of similarity or difference. The meeting between Roland and Caisley in the final scene is particularly ambivalent. The four sections do not add up to a holistic view of life; instead each section antimediates the others to privilege uncertainty and potential. For all of the interest in the neutral zones and the fixed points that obsess the characters, Modiano's own method privileges connectivity and openness. When Caisley unexpectedly lays his hand on Choureau's shoulder, for example, the sterility of the featureless apartment dissolves. The most intimate moments between Louki and Roland occur in the prefurnished rooms of the Parisian neutral zones. And there is the café itself: the café *in* which the events of the story unfold and are retold.

The modernist question of alienation versus connection plays heavily in Modiano's work. So, too, do the issues of apathy and agency. These binaries largely determine the environmental control and determination that runs throughout the Parisian milieu. At the same time, the text is also aware of the social transformations that Guy Debord would outline in his 1967 work *The Society of the Spectacle.* Louki's juvenile vagrancy, for example, begins with trips to the Ciné Mexico at the end of her street.[36] Caisley notes the appropriateness of the name, as it suggests the desire "to travel, to run away, to disappear."[37] This escapism is certainly a major factor of Louki's vagrancy, yet in her own narration of filmic experience, the spectacular effects of the cinema are more complicated: "I paid no interest to the plot, only the scenery interested me. Once I was back outside, I was left with a curious amalgam of Arizona and the boulevard de Clichy in my head. The coloring of the illuminated signs and the neon were the same as that in the film—orange, emerald green, midnight blue, sandy yellow—colors that were too violent and gave me the feeling I was still in the movie or in a dream."[38] In her exclusive interest in scenery, Louki focuses on the environmental aesthetics of film. Film presents the unique experience of witnessing the three-dimensional world on a single screen, collapsing the physical space of the theater while at the same time inviting the viewer to imagine inhabiting foreign locations.

While the ceremony of viewing a film—the dimming and raising of the lights—serves to differentiate the filmic world from the real world, for Louki the experience is not so simple. The intermingling of the film with the boulevard that she experiences when she reemerges onto the streets of Paris is a hallmark of Debord's spectacle, which is never simply contained by its technological apparatuses but always signals the permeability between the real and the virtual. Louki's insight into what Debord terms the "reciprocal alienation" of the spectacle allows her to begin changing her relationship with her environment. Her experience in the theater leads to the détournement of her perception of everyday life and contains the seeds of a developing environmental aesthetic that culminates in the summiting of Montmartre. Instead of escapism or the apathy and passivity that are hallmarks of a complacent consumer culture, Louki's engagement with spectacle allows her to recognize the surreal and hyperreal elements that influence her everyday life.

This leads back to the question of what influence environmental aesthetics can have on the physical world. Scholars question the place of environmentally minded works of art and literature for advancing activist agendas. Malcolm Miles, for example, notes the facility with which global capitalism can incorporate environmental aesthetics to make its products and services more desirable under the banner of green marketing. At the same time, he differentiates a "critical aesthetics which exposes contradictions *and* offers access to moments of wonder and glimpses of another world. But that world is *this* world transformed either in an imagined alternative scenario or practically, if as yet marginally, within the present ordering of society."[39] The trajectory of Louki's environmental aesthetics—from the Ciné Mexico to the top of Montmartre—seems to fall into this category of a critical aesthetics, full of contradictions that nevertheless allow her to question her surrounding world and the changes taking place within it. This associative and imaginative mode provides an alternative to the cartographic strategies of Bowing's fixed points and Roland's neutral zones.

Ephemeral Utopias

If Louki's death provides a rather stark end to her growing environmental awareness, what aesthetic responses are viable? As the architectural and technological sprawl of civilization envelops the passive spectator, can détournement take on an ecological dimension that might lead to new behaviors toward and understandings of current environmental issues? One paradox is that technology mediates knowledge of climate change. As Morton writes, "Machination ruins Earth and its lifeforms, yet it supplies the equipment necessary for human seeing at geotemporal scales sufficient for ecological awareness."[40] Ever more sophisticated climate models based on satellite imagery and complex algorithms aim to create global simulations of the future. A rise in disaster entertainment employs computer-generated images to present climate change in ever more grandiose terms.

The spectacular aspects of environmentalism may lead to undesirable ideologies. Alain Badiou suggests as much by identifying a millenarianism within environmentalist rhetoric that posits ultimate destruction. "These visions of final catastrophe," he writes, "tend to mobilize everyone to conserve the planet just as it is, with its temperature, its pleasant 'environment' for the petit-bourgeoisie of rich countries, and its predators of globalized capitalism."[41] While environmental efforts can reach a larger audience than ever before, they are not immune to the paralyzing effects of spectacle. One might be encouraged, then, to question whether the spectacle surrounding the Anthropocene serves to generate change or, conversely, to encourage viewers to further detach from their actual physical surroundings. There is an opportunity for innovative aesthetics to challenge global apathy in the face of mounting media-induced climate change fatigue, but such aesthetics will have to contend with the material forces that drive seemingly unstoppable feedback loops.

With the civil unrest in France in May of 1968, what had been abstract and theoretical for the Situationists became violently real. With these events détournement ceases to be a playful commons and instead becomes an aggressive arena, providing a watershed moment from which to reconsider the relationship between theory and praxis. The group had already weighed in on similar demonstrations occurring in Egypt as well as the 1965 Watts rebellion in California. Taking place on a particularly hot summer night when rolling blackouts removed citizens from their air-conditioned apartments, the Watts protest unleashed escalating racial and economic tensions that had been building for decades. Resisting the media portrayal of demonstrators as "looters" and "criminals," the SI saw in the Watts rebellion a *natural* response to an *unnatural* life brought about through capitalist exploitation. Given that the events took place in the greater Los Angeles area, a stone's throw from Hollywood, "pole of the Global spectacle," the SI was quick to incorporate these events into their own narrative, claiming that "a revolt against the spectacle—even if limited to a single district such as Watts—calls *everything* into question because it is a human protest against a dehumanized life, a protest of *real individuals* against their separation from a community that could fulfill their *true human and social nature* and transcend the spectacle."[42] Rebellion, revolt, and revolution are interesting test cases for theories of the spectacle precisely because such demonstrations are also a spectacular break from the everyday. As such the SI calls out a need for a new self-reflexivity among revolutionary movements to avoid cooption by media and other interests.

The question remains: What will be the role of revolt, rebellion, and revolution in environmentalist action? Ecologically minded activists are notoriously peaceful, although certainly not spectacle averse. From chaining oneself to a tree in an old-growth forest to sailing across the Atlantic Ocean using solar power, environmentalists have not shied away from media-baiting stands. More aggressive direct action such as ramming a whaling boat or blowing up a pipeline provides an outlet for anger and frustration but may backfire if not met with reasonable alternatives. This leads some to reluctantly embrace a more market-driven approach. The recent surge in solar power has not been occasioned by any drastic actions or clever campaigns; instead, the low price of 5 cents/kW hour is what drives the adoption of this more environmentally friendly technology.

While the Situationist International disbanded in 1972, the theoretical and artistic interventions made by the group continued to influence discussions of space and environment. The texts and projects presented in the journal *Utopie,* a "review of urban sociology" published by Éditions Anthropos in Paris from 1967 to 1978 showcase the afterlife of spectacular thought and lived experience. The group presents the thrust of utopia as a dialectical one, seeking to exist in the "uncrossed interval between theory and praxis."[43] Informed by theoretical underpinnings, participants may free the "social possibilities" confined within the urban environment and put "their revolutionary potentiality into play."[44] Utopia for the collective is not a destination but a constant process of call and response toward liberation.

Most convincing in the early issues of *Utopie* are the new aesthetic explorations. René Lourau outlines an environmental aesthetic that does not seek to tame or exert control over surroundings but, rather, questions how art happens when, "in the presence of what we call the external world, representations take shape within us."[45] Constant's 1968 collection of *Labyrismen* presents a series of figures who appear to be navigating this contradiction between internal sensation and external stimuli (see fig. 11). The series is at once expansive and claustrophobic, as figures appear both trapped and at the precipice of unlimited unfolding paths. Jean-Paul Jungmann's article "Paper Cities" explores the limitations of representing space in a two-dimensional medium. The essay cleverly questions whether the visual medium of "graphic writing" in fact hampers "visions" for the future of urbanism. The pressures of building competitions, marketing, journalism, and other media outlets lead to a visual language that has little incentive to consider social issues.[46] It's worth noting that Jungmann's writing style itself is highly graphic, as each page presents a collage of quotations, commentary, drawings, and photographs. Nevertheless, the paper city presents a rift between reality and deceptive forms of representation and marketing.

These ideas take a more environmentalist turn in Jean Aubert's "Becoming Outdated" and Jean Baudrillard's "The Ephemeral," which consider both the perils and promises of a throwaway society. For Aubert,

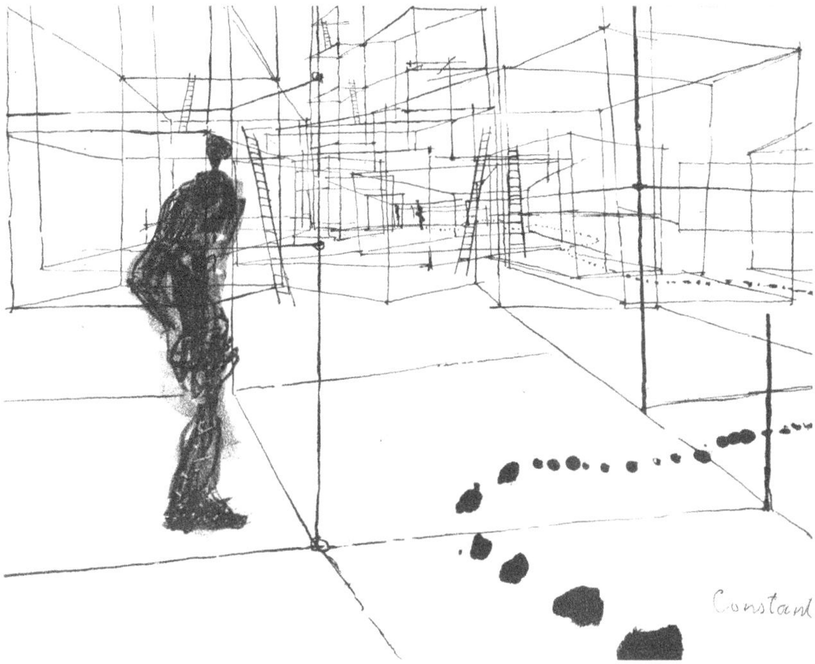

FIG. 11. Untitled, Constant, from *Labyrismen*, 1968. (Digital image © The Museum of Modern Art/licensed by SCALA, Art Resource, NY, with permission from the Artists Rights Society)

ephemerality is the future, and durability is the past. He imagines future cities as great galleries decorated with ephemeral works that vanish and reemerge in new permutations. He revels in the potential of the ephemeral for transforming the streets: "Color is everywhere; light, too. Day and night, color and light reecreate the forms in never-ending variations, created by chance or sometimes by a specific will, that of the Esthetes who are past masters in the composition of the most amazing urban landscapes, and at the service of industrial firms using their talent to commercial ends. Curtains of steam, curtains of insulating air, translucent paper, synthetic materials are deployed under the control of the light to create a continuous spectacle."[47] In Aubert's fantasy, art, engineering, and industry all converge under the benevolent vision of Wildean "esthetes" who curate a world of perpetual invention. Boredom ceases to exist and discomfort is a fleeting suggestion before a new cycle of experience begins.

Baudrillard, however, brings this fantasy back to earth by emphasizing the social strictures at play in ephemerality. While he agrees in the coming future of the ephemeral, he emphasizes that it is one that will be primarily for the benefit of the bourgeois and upper classes, those who have enjoyed centuries of security and permanence through the solidity of private property.[48] It is this view toward social striation and inequality that grounds Baudrillard's critiques of media and environment. Like the Situationists before them, Utopie was highly skeptical of mainstream environmental agendas. Perhaps as close as they get to an environmental statement is refusing to attend various international conferences, reducing their carbon footprint by sending papers to be read in lieu of attending in person.

This move is thrown into starkest relief at the 1970 conference on environment where the Utopiens not only refuse to attend but send a paper denouncing environmental movements. The unsigned "The Mystique of Environment" does not mince words in claiming that environmental concerns divert revolutionary energy away from social agendas, claiming that "the environment issue is a fallout of May 1968, more precisely, a fallout of the failure of the May revolution. Ideology, which the political power tries to divert onto rivers and national parks, could happen in the street."[49] In the essay's argument, issues such as pollution, scarcity, and ecosystem collapse are mere environmental bogeymen that seek to increase social stratification. The myth of human unity with marvelous nature simply reinforces the "holy union of the ruling classes of the rich nations."[50] The designation of the Anthropocene faces similar critiques for its obfuscation of social inequality in the "human epoch." Environmental justice movements importantly insist on the inextricable link between the social and the ecological. Yet the question of which environmental issues gain media attention and why remains crucial for critique in the twenty-first century.

Also in 1970, Baudrillard published a similar polemic taking aim at media. In his "Requiem for the Media," he critiques leftist media theorists who denounce media culture but dream of hijacking these very same technologies in the service of revolution. Baudrillard again points to May 1968, where media appeared to amplify and expand the rebellion. The fact that "the media maintained their form (amid the disruption of their contents)" belies the inescapable problematic that media are "inexorably in solidarity with the power structure."[51] There is no détournement of media just as there is no radical ecology that can revolutionize humanity's relationship with nature. Both programs falter in the realm of the social. The question of media's democratization versus monopolization becomes central as more and more people participate in a global web of social media interactions.

While the revolutionary aesthetics pursued by groups such as Utopie and Archigram did change spatial practice in profound and important ways, for the most part it ended quite shy of the completely remodeled urban experience imagined in the more fantastical visions of Aubert and others. Instead, the ephemeral and malleable conditions so desired in the physical realm ended up finding purchase in the digital realm and the increasingly sophisticated world of art installations. The blending of art and technology is alive and well in the works of artists such as James Turrell, Olafur Eliasson, Andrea Polli, and Daan Roosegaarde, among others, who create large-scale situations that call attention to issues on a planetary scale.

Turrell expertly shapes texture and light to create gyroscopic art spaces that challenge a viewer's sense of proprioception and planetary place. His ongoing project at Roden Crater presents a celestial observatory meets doomsday bunker to make the viewer both out of body and grounded in mantle. Eliasson's iconic *The Weather Project* installed in Tate Modern in 2003 presented a hazy microclimate of light and fog which invited viewers to interact in the large open space of the former power station. The more pointed *Ice Watch* of 2015 placed large chunks of Greenland's ice cap in the Pantheon of Paris to slowly melt while world leaders gathered for the Global Climate summit. His 2021 work *Life* opened Fondation Beyeler to the elements creating a living ecosystem accessible to visitors 24/7. The artificially bright green combined biology with technology to create a work both reassuring in its naturalness and unsettling in its vibrancy.

Polli's piece *Particle Falls* projected real-time air-quality information in the form of a waterfall in Pittsburgh in 2014. A more recent project, *N-Point,* presents time-lapse video of the Arctic with an immersive sound installation of weather at the North Pole. Daan Roosegaarde's *Waterlicht* employs LED, lasers, and atmospheric fog to create a "virtual flood" that can be installed in important social spaces such as town squares or city halls. Visitors can reimagine these sights under water while they, too, walk around and socialize under the inundation.

Sophisticated art installations face the paradox of placing people in highly constructed worlds to achieve insights into physical reality. On the other hand, these works can, importantly, take art outside of the already artificial museum to meet people on the street. Encountering media in this fashion can hijack a person's day and redirect their attention toward a contemplation of the planetary. As the next two chapters will argue, such physical moments of social connection are proving more crucial as people increasingly pursue digital mediation in the virtual environments provided by computers, mobile devices, and data networks.

PART III

THE GOLDEN AGE OF (DIS)CONTENT

5

Cyber Saturnalia

> Everything real grew misty, and everything misty was without boundaries. Onslaughts on the present, broad and shapeless feelings, nothing natural, everything constructed out of the blue, everything the precise opposite of what is and what ought to be.
>
> —KARL MARX, letter to his father, Heinrich, November 10, 1837

NOW THINGS are really speeding up. The world is changing overnight, at the pace of a screen refresh at 120 hz. Visual technology has become more sharp, too precise for what the human eye can appreciate. For early media devices such as the zoetrope and even the film reel, speed was a crucial factor, but not in an absolute sense. Instead, it was the precise calibration of speed, a certain number of frames per second that allowed the magical illusion of movement to take place. Run too quickly, the effect backfired. But for the hyperworld of computing and internet connectivity, speed is pursued as an absolute: more processors, more bandwidth. If there is a limit to the onrush it has not yet been found.

There is a certain seduction to the online energy equation. Gilbert Simondon describes this as the low-wattage technologies of digital mediation. Better to teleconference than drive. Why fly to an exotic locale when you can explore the same place in virtual reality? Twenty-four-seven connectivity leads to constant access to all corners of the Earth and all imaginable forms of content.

A retreat into cyberspace holds its own perils, beckoning in a similar fashion as the pastoral arcadia: enter here where everything is clean, glossy, smooth, and lightly glowing. With this subtle mediation, a new layer of existence emerges. Sherry Turkle writes, "Our new devices provide space for the emergence of a new state of the self, itself, split between the screen and the physical real, wired into existence through technology."[1] Antimediation becomes harder to achieve in a world perpetually plugged in. Still, writers and new media artists wade into these spectral waters to disrupt the Saturnal gobbling of time and interpersonal relationships.

Common Boundaries

Jacques Derrida famously founded his deconstruction on the sly recognition of the structurality of structure.[2] With the contemporary technomedia landscape, however, the structure is—at one level—all too evident: the screens in front of faces, the light switch that powers the room, the gas station that pumps gallons in minutes. Further, as Simondon argues, the expansion out of city centers via transportation lines, electrical grids, and information cables marks a new interface with key nodes of the landscape. This process complicates older conceptions of the human conquest of the elements or the industrial destruction of nature; instead, "it is the natural structures themselves that serve as the attachment point for the network that is being developed."[3] This integrated system gives rise to a "technical mentality," which enmeshes nature and technology, materiality and thought, individual and community.

At the same time, most interaction is spent not with structure itself but rather with an interface that mediates between structure and user. The interface is the forward-facing, friendly, and accessible aspect of technology. The move to graphic operating systems brought about by Apple and Microsoft accelerated the widescale adoption of computing products. The interface, however, is not merely a concession to human disposition. It works both ways. The interface is a shared boundary that allows disparate components to overlap, exchange information, and, perhaps most significantly, *act* on one another.

One of the earliest uses of "interface" as a verb comes from David Packard. He of Hewlett-Packard fame, his 1969 article "Publishing Scholarly Compilations by Computer" in the journal *Computers and the Humanities* surveys the nascent field of computer-assisted publishing. Packard notes the arrival of machines that are "entirely electronic, the characters are generated digitally, with no physical prototype characters."[4] He references a Belgian professor who "interfaces" his Selectric typewriter to his local computer to publish the journal *Revue* as well as indices to classical texts in a flexible but "very, very slow" fashion.[5]

The interplay between user and object takes the form of technogenesis, where, in the words of Katherine Hayles, "humans and technics have coevolved together."[6] Hayles draws on epigenetic models of biological changes to rethink technological development, "Contemporary technogenesis is about adaptation, the fit between organisms and their environments, recognizing that both sides of the engagement (humans and technologies) are undergoing coordinated transformations."[7] Astute as this observation is, there is no attempt by Hayles to engage larger *environmental* transformations spiraling out of control of either human or technological agency. Environment is too often treated as apparatus rather than as an unpredictable force reeling from technogenesis itself. If humans have lost control of the physical environment, a new faith arises in the ability to control virtual spaces.

How to navigate these dizzying variables? How to *mediate* between human, technology, and an increasingly unpredictable world? "Without the aid of the 'artificial horizon' of multi-media," writes Paul Virilio, "there is no possible way of negotiating the electronic ether of globalization."[8] Is there an artificial horizon for navigating environmental crises? Whereas technology begins as a desire to master nature, Walter Benjamin sees in the art of technological reproduction an "interplay between nature and humanity" that is based in "play," but a play that prepares the subject for new forms of production unleashed by these very modes of technological innovation.[9]

If play can be coopted into productivity, others seek to reclaim play as an agentive force for change. Alenda Chang looks to inventive video game designs that can "create entirely new sets of relations outside of those based on dominance or manipulation" and model "different forms of our individual and collective environmental agency."[10] Virtual environments can generate different relationships to real environments, and events such as the Green Game Jam seek to harness the enthusiasm of gamers to increase awareness of and raise funds for wildlife conservation. Gamers, in turn, can reconsider their diets, support sustainable brands, and volunteer their time for habitat cleanup.[11]

This emphasis on connecting the virtual world to the physical is indicative of a shift where technology no longer wants to be a separate space but prefers to be seamlessly integrated into everyday life. To understand the normalization of the interchange between people and their devices, Stéphane Vial explores a phenomenology of digital mediation and its profound effects on what he calls "ontophany," the way in which beings appear. Living in the information age means encountering new "digital beings." These encounters in turn change what it means to perceive and appear in the material world.[12] Vial's is a more technophilic interpretation of the digital age. To be a digital native is not some escape from reality or sign of social awkwardness, just the opposite: the digital native is fully present in the contemporary world and in touch with its technological moment.

On the encounter with digital beings, Ted Chiang's story "The Lifecycle of Software Objects" reveals just how potent connections may become between the human species and a coming evolution of digital organisms. The story focuses on Digients, artificially intelligent companions that inhabit the realm of Data Earth. Fitted with a neuroblast engine, the Digients are capable of learning and developing over time. They are also able to inhabit mechanical bodies in the real world to further interact with their caretakers.

Just as physical environments undergo changes of technogenesis, Chiang describes the rise and fall of online worlds as similarly subject to capitalist trends. A longtime stalwart of cyberspace, Data Earth succumbs to a rival: "Currents within the mantle of the marketplace change, and virtual worlds undergo tectonic shifts in response: a new platform called Real Space, implemented using the latest distributed-processing architecture, becomes the hotspot of digital terrain formation. . . . Data Earth has long been a fixture in the universe of virtual worlds, resistant to growth spurts or sharp downturns, but now its topography begins to prod; one by one, its virtual landmasses disappear like real islands, vanishing beneath a rising tide of consumer indifference."[13] Chiang's geologic language is metaphoric, but it also points to the flow of signifiers of human knowledge that find new signs to attach themselves to in the digital world.

After a flu pandemic throws the economy into recession (speculative fiction at its finest), Data Earth is subsumed by Real Space, throwing the life of the Digients into peril. Digital extinction is treated with the same gravitas as actual extinction. At the same time, there is little interest in saving the Digients from those other than the small group of devoted caretakers. This lack of concern qualifies virtual attachment. While it is possible for profoundly real connection, there is not necessarily a larger understanding of digital relationships in the same way that humans are able to extend empathy to one another, animals, and indeed the entire planet.

What connects both digital and analog life, for Chiang, is experience. Thinking about the time spent with her Digient, Jax, Ana cannot help but conclude that it is not time wasted, but time lived:

> Experience isn't merely the best teacher; it's the only teacher. If she's learned anything raising Jax, it's that there are no shortcuts; if you want to create the common sense that comes from twenty years of being in the world, you need to devote twenty years to the task. You can't assemble an equivalent collection of heuristics in less time; experience is algorithmically incompressible.
>
> And even though it's possible to take a snapshot of all that experience and duplicate it ad infinitum, even though it's possible to sell copies cheaply or give them away for free, each of the resulting digients would still have lived a lifetime. Each one would have once seen the world with new eyes, have had hopes fulfilled and hopes dashed, have learned how it felt to tell a lie and how it felt to be told one.[14]

Chiang builds time into this story. It is a long story. Time is discussed matter-of-factly in realist style. Sections often begin with "Another year passes," "Two months go by," and other measures of duration. As the characters age and their relationships change over time, the Digients also struggle to find a way to gain autonomy and take control of their own destinies.

Salvation for the Digients comes from an offer from Binary Desire to port the Neuroblast engine over to Real Space to develop Digients to be romantic partners for human beings. This leads to an existential crisis for the group: Even if done in an ethical fashion and with the consent of the Digients themselves, can one justify altering a digital being's existence in such a fashion? In a particularly heated moment, Ana blurts out "It's not real," a statement she immediately realizes calls attention to the elephant in the room. While the next chapter on "Digital Twins" will discuss artificial intelligence and virtual reality in greater detail, Chiang's story suggests that digital relationships *are* real. Time spent with technology is as meaningful as time spent with plants, animals, and humans. Ana's relationship with her Digient is not misplaced desire for children or some more holistic connection with the real. Jax has become her reality.

Attention Ecologies

With the proliferation of personal computers, mobile devices, video game consoles, wireless internet, live streaming, social networks, fitness trackers, self-driving automobiles, and AI assistants, the ruling logic and expectations of the society of the spectacle begin to break down. The roles of actor, director, and audience begin to blur. The strict binary of active versus passive begins to deconstruct. Similarly, the distinction between extension and compression no longer seems so straightforward. Ecology is one metaphor to think through an increasingly tangled web of technological being; however, it fails to adequately model the intersections between the real and the virtual, the material and the imaginary.

As the geophysical effects of humans increase alongside more technology, more convincingly real simulations of experience, more mediation between humans and their environments, the Mediacene threatens to displace traditional notions of nature and environment completely. The global expansion of screens and "terminals of tele-action" lead increasingly to what Bernard Stiegler terms "the industrial temporalization of consciousness."[15] Moreover, from an emissions standpoint, this is perhaps a desirable development. As Simondon notes, "Electronics and telecommunications use reduced tonnages, moderate energies, dimensions that are not crushing."[16] Exploring the world in virtual reality, in other words, carries a much lighter carbon footprint than making use of the industrial assemblages of cars and airplanes. Others, such as Yves Citton, however, question just how moderate these networks truly are, citing among other factors, "the electricity consumed by the servers supplying the web, the huge increase in toxic waste caused by the built-in obsolescence of our personal computers and mobile phones, the increasing proportion of household budgets taken up by connection costs, the debt spirals engendered by the ease of online purchasing."[17] These light tonnages, in short, add up to heavy environmental tolls. Beyond simply focusing on the material aspect of networks, Citton's work helps to shift the attention to, well, *attention*. An attention to attention is one way to begin to bridge the gap between the material and the metaphysical in discussions of mediation.

There is an economy of attention as well as an ecology of attention. The economy of attention can be measured, parceled, and monetized. Arthur C. Nielsen pioneered audience ratings for radio in the 1930s and television in the 1950s, but the ecology of attention is slightly more amorphous and unpredictable. For Citton, attention is not passive but inherently interactive; he calls it the "essential mediator" for surviving and thriving in an environment.[18] Attention itself is a mediator, but it also exists within a multilayered system of mediations from biology, sociology, psychology, geopolitics, and so on. Citton considers the mass of information communications to form a "mediasphere," echoing the noosphere, blanketing the globe not unlike a layer of greenhouse gasses. Within this mediasphere there exist any number of "attentional ecosystems" that direct group attention through aesthetic and affective resonances.[19] Here is the crux of an attention ecology: alone, the viewer becomes alienated and isolated, but, together, group attention can bring focus to not only "act" in the environment but change and intervene in existing structures of mediation.

Working against the collective agency of group attention is *distraction.* In his work *Infinite Distraction,* Dominic Pettibon theorizes distraction as an end in itself. Instead of being used as a decoy or an attempt to hide something else, distraction is now built into media coverage: "Deliberately framed representations of events are *themselves* used to obscure and muffle those very same events."[20] Distraction may tap into physiology and bodily reflexes to create antimediation and divert attention away from mainstream narration to reveal a "glimpse of reality." At the same time, as Pettibon warns, such insights are themselves easily subjected to "re-distraction," to new avenues of separation and alienation so that the new, "interactive spectacle" ensures that potential allies never meet or share the same moment.[21] Distraction is both the means and ends for disrupting the "attention economy" into a proliferation of "different attention *ecologies,* each with its own ecosystem and microclimate."[22] On the one hand, this is a good thing and part of modernization's longer project of liberalizing communities to embrace diversity and a multitude of responses to the human condition. But, on the other hand, when considering the potential for these attention ecologies to come together to confront the monolith of the Anthropocene, it is easy to see the challenges to forming any sort of consensus or organized action.

One solution, if there is such a thing in the intricate and unfolding complications of contemporary life, may be to embrace the small toward microprogressions. If a community wants to band together online to protect a specific forest or species, more power to them. If another wants to focus on environmental justice issues related to toxic air around oil refineries, have at it. Small gains are desperately needed, even if at a certain level these local battles may distract from the larger war of anthropogenic change. As soon as attention is fixed on any one issue, object, or person, a whole assemblage of forces work to distract that attention into different avenues of consumption. Media often arouse an emotional *response* and that only: through the screen one is able to *feel,* to be excited, sad, to cry, even, before ultimately returning to everyday life without a deeper reevaluation of behaviors and activities.

This short-term affective response acts as a therapeutic release for a fundamentally unsolvable neurosis. As Wilde puts it, "We weep, but we are not wounded."[23] In Aristotle's classic formulation, the emotions aroused by tragedy's mimesis of life can provide catharsis, a cleansing that can lay the path for renewal, restoration, or even transformation into something new or previously unimagined. For Aristotle and Wilde, emotional storage builds through the day-to-day slog with physical reality, a slow and methodical storing up of unsatisfied *energy.* To tap into these reserves, to release the emotional pressure is the goal of art, theater, and literature.

Today's mediascape encourages the opposite: the practice of "binge-watching" provides the viewer with an endless stream of highs and lows that gradually lose emotional weight. The catharsis, or the purge, now becomes a release from content and a return to real life: moments of death, disaster, or intervention. The new light and associative environmental aesthetics of immersion and ambience also seem easily coopted by the larger thrust of anthropogenic attention capitalism. Immerse yourself in the virtual world! Experience the sensations of nature itself! Go outside and catch Pokémon on your phone!

Just as in Manfredo Tafuri's account, where urban naturalism arose alongside the picturesque, the new environmental aesthetics of vibrant matter, object-oriented ontology, and hyperobjects arrive alongside humanity's reflection on itself as a world-shaping force. If storms, forest fires, droughts, and rising sea levels hail an increasingly *détourned* nature, the media fanfare around such events keeps up apace. If augmented nature takes down the power grid momentarily, nothing becomes more imperative than restoring the electricity that feeds the virtual web and controlled living environment. This is not to invite a new apathy in the face of environmental issues that snowball out of human control, but to recognize new challenges to aesthetic projects that encourage humans to interrogate their global impact.

The interface of technological devices not only expands the potential for connection but also adds a layer of complexity to human collaboration. Jean Baudrillard theorizes a networked systems of screens through the trope of *obscenity.* His 1987 work *L'autre par lui-même* (*The Ecstasy of Communication*) arrives not only as a pivot into his later works, but also as an attempt to move beyond the tropes of spectacle and surveillance pioneered by his compatriots, Guy Debord and Michel Foucault, respectively. The rise of ecological movements in the 1960s and 70s creates a shift from a semiology of objects toward an analysis of networked mediation. The vehicle by which Baudrillard moves from one to the other is *energy:* a renewed focus on consumption and expenditure. It is this baseline currency that drives the contemporary age and displaces the decorum of spectacle for an obscene openness. As Baudrillard explains, "The consumer society was lived under the sign of alienation; it was a society of the spectacle—but at least there was spectacle, and the spectacle even if alienated, is never obscene. Obscenity begins when there is no more spectacle, no more stage, no more theatre, no more illusion, when everything becomes immediately transparent, visible, exposed in the raw and inexorable light of information and communication."[24] This, for Baudrillard, signals the ecstasy of communication: a world marked by the pornographic, polluted, and promiscuous exposure of the hypervisible.

The obscene functions as a certain blurring of εἶδος (form, image, shape) and εἴδωλον (phantom, fancy, ghost). As such, ecological relationships become amorphous as well. In the Baudrillardian obscene, the human does not live in an ecological relationship with the world so much as one of pure exposure and conductivity, a schizophrenic state of narcissistic dispersal. No longer the architect carefully managing blueprints and hovering above the construction of the planet, humans have become as a spider stuck in its own web. The way out of this trap for Baudrillard is through an endless slog of transcendence, an exploration of mutability and metamorphosis.

Humans strive to escape their biology and become a different species, yet environmental justice presents the imperative to consider vast differences among the species homo sapiens itself. There is a model for this kind of transcendence seeking *within* society. As much as humans may daydream about the lives of birds or dolphins, such fantasies are surely dwarfed by the overwhelming desire to transcend into the lives of other humans. To see how the other side lives, to become a voyeur or tourist, to walk a mile in another's shoes. Instead of the metamorphosis pursued by Ovid and others, might environmental aesthetics benefit from considering the Saturnalia?

Saturnalian Praxis

Civilization has entered a new golden age. Under the beneficent reign of Saturn, miracles are coaxed out of the soil. A click of a button, a swipe of the finger, and one can summon food, clothing, a new relationship, an autonomously driven vehicle. Venture capital pumps through the cities, valuing unprofitable companies at incredulous levels. In the cyber world, one can choose and customize an avatar to be anyone, anything, any combination of exotic or banal characteristics. One can simulate any fantasy, find niche communities around the most obscure of topics, connect with people all over the world without ever leaving the comfort of home.

What happens in a world of instant gratification, social media mobs, and technological abundance? The reign of Saturn ended in catastrophe. An aging ruler desperate to keep his grasp on power, Saturn devoured his own children to extend his rule, a crime gruesomely captured in Francisco Goya's famed portrait (see fig. 12). In Goya's painting, Saturn desperately gnaws on his progeny to avoid fading into the pure black background of oblivion. While crude, this metaphor is constantly invoked in environmentalist rhetoric: think of the children, the coming generations, the injustices they will inherit are equally devastating.

Against this vision of deranged consumption, there are other interpretations of Saturn that give the Titan the name Sterculius, the god of manure, to reflect the period of fertilization. To rethink of Saturn in terms of compost, organic biomass, and the recycling of resources may lead to a more productive, lasting, and sustainable approach. No longer a golden age but an age of dung. The Roman historian Macrobius recounts as much in his long work *Saturnalia*. Set during the festivities of the winter solstice, the numerous chapters expound on great thinkers from history in the name of celebrating renewal and social harmony. Writing in the tradition of natural history established by Pliny the Elder, he details a process of producing knowledge akin to the distillation of pollen into honey performed by a hive of bees.

Nature, for Macrobius, is an aggregator, wherein data is accumulated and eventually transformed into new energies. This is evident in the process of digestion, a metaphor which Macrobius extends to explain the metabolism of the mind: "We must see to it that we do not allow what we have absorbed to remain unchanged and thus fail to assimilate it, but we must, as it were, digest it. To do otherwise is to feed the memory,

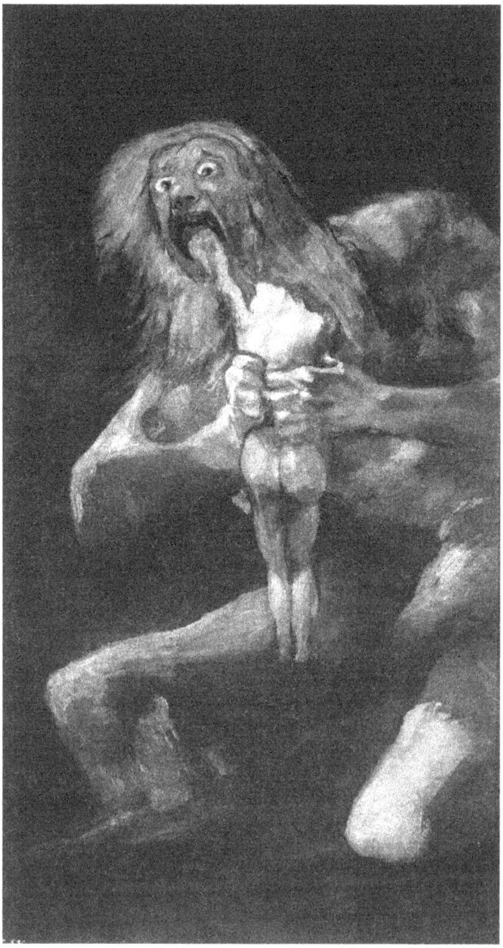

FIG. 12. *Saturn Devouring His Son,* Francisco Goya, 1823. (Museo Nacional del Prado)

not the mind."[25] The distinction between memory and mind here is instructive. Unlike the negative consumption of Saturn devouring his children, a practice of self-preservation aimed at one's own memory, the Macrobian process of consumption is one that empowers others, inviting as many voices as possible to join in a chorus of learned expression. Stiegler identifies a similar gobbling of global memory taking place as industrialization enters the digital age: "In order to face the new requirements resulting from the process of permanent innovation at the heart of the Industrial revolution, information systems have become global, the result having been—through the development of the telegraph, telephone, photography, cinema, radio broadcasting, television, and the information technology whose emergence is currently taking place—that global memory has itself finally been subsumed into an industrialization directly affecting our psychic processes and collective identifications and differentiations; that is, individuation itself."[26] Too much memory may lead to paralysis, but too little threatens to lead to disassociation, loss of self, and dissolution of community. The apotheosis of this process is Jonathan Crary's world of 24/7 work and activity, a system that is always on, always present, and, as a result, full of depletion.[27]

Breaking through the industrialized assault on memory and mind in the information era requires antimediation to disrupt the constant stream of consumption. The Saturnalia was precisely such a pause in the normal flow of life to provide time to reevaluate human activity. At the same time, this state of exception was sanctioned to ultimately reinstate the ruling order to greater effect. The question remains: To what extent can antimediation reroute technogenesis? The form of Macrobius's treatise takes on something of a Platonic Symposium 2.0. Set during the "rest days" of the Saturnalia, various nobles, politicians, and scholars meet to feast and hold discussions on ethics and the liberal arts. Unlike the predetermined agenda of Socrates, however, Macrobius seeks to let the voices have it out, to reach forms of harmony and discord from which a higher sense of knowledge will emerge. This is also the promise of social media systems that privilege user content such as YouTube, Instagram, Twitter, TikTok, and Facebook: now anyone may pick up electronic tools to join in the symposium, create content, and shape their own community of subscribers.

With an active user base nearing three billion people, YouTube is crowdsourcing creativity to develop content for all walks of life. The proliferation of videos aimed at exploring the wilderness, bushcraft, sustainability, and self-sufficiency reveal the potential for these platforms to create a robust dialogue around environmental issues. Several series revisit the Thoreauvian imperative to live simply in harmony with the land. Often absent from such videos is precisely the technological apparatus that makes it all possible: the cameras, the computers for editing, the cable networks that upload the video, and the vast infrastructure that delivers it to devices around the world. As a platform ruled by views and engagement, YouTube is an interesting test case for those who would thrive in the gilded age of media: a way to engage in creativity and laudable causes while still being rewarded by the capitalist monetization of user attention to sell advertising.

Computer algorithms seek to aggregate data to simulate human decision-making processes and arrive at a system of machine learning that considers memory, stimulus, and search. But too often contemporary algorithms lead toward a process of reduction, feeding precisely on the "memory" of stored inputs rather than the "mind" of associative potential. One's feed, rather than flourishing outward, tends to narrow and collapse. This is seen most strikingly in the "targeted" advertising of social media sites, where, seconds after mentioning a particular interest or issue, one is invited to click a link to get more information and purchase related products.

What might it mean to refashion the cyber Saturnalia away from an endless loop of binge-watching, orgiastic indulgence, and expenditure toward the regenerative Sterculian model? The planned obsolescence of technology and consumer thirst for the next device do not help. While there is an interest in retro-tech as fetish objects and dumb phones that put function over addiction, what may be of more help in the information age is analog tech: the old mediums of books and paintings may provide the intellectual dung to change and reinvigorate the constant, banal hum of low-wattage processors. In addition to "old media," electronic literature, comparative media art, and creative projects in virtual and augmented reality provide helpful digital manure to replenish a commons depleted by algorithmic echo chambers that deaden human potential.

Weathering the Real

Just as poets and novelists grappled with the emergence of radio, film, television, and other media technologies in the twentieth century, writers in the twenty-first century are increasingly engaging the mediated realm of the internet. In some ways, this has been a more natural fit, as the Web began and, in many aspects, remains a predominantly textual experience. Today there are more readers and writers than ever before in human history, and the development of language is arguably changing at its fastest rate since the invention of the printing press. Unlike the symposium or saturnalia that occurred once a year, the digital arena presents constant access to debate, new ideas, (dis)information, and a growing chorus of voices.

Two recent works that integrate a shorter, associative writing style to cater to the multitasking attention spans of the online era are Patricia Lockwood's *No One Is Talking About This* and Jenny Offill's *Weather*. Lockwood tells the story of a woman who has become a viral internet voice on "The Portal," stemming from a random musing, "Can a dog be twins?" The narrator compares the stream of collective consciousness to her modernist forebears James Joyce and Virginia Woolf. Joyce's metempsychosis, the transmigration of souls, becomes, in the digital age, a corporeal dissolving. The portal provides both a dizzying sense of connection as well as profound isolation: "It was in this place where we were on the verge of losing our bodies that bodies became the most important. . . . In this place of the great melting . . . you were zoomed in on the grain, you were out in space, it was the brotherhood of man, and in some ways you had never been flung further from each other. You zoomed in and zoomed in on that warm grain until it looked like the coldness of the moon."[28] The portal is omnivorous; it brings people together through mutual hate, niche interest, comedy, sexual proclivity, and slang. At one point the online activity of humans is compared to a species of cephalopod releasing puffs of poison and black ink clouds on the ocean floor.

While climate change is not the main interest of the book, it does feature in a few key sections. The narrator details a performative piece of activism where a man walks barefoot across the country, posting a photo of his feet everyday:

> There was a poet who was walking across America barefoot to raise awareness of climate change—how was this supposed to work, exactly? Yet she breathed to herself the words *climate change* whenever his name stepped toward her in the portal. He posted a new picture of his feet every day, so that she saw the innocent blisters spread and break, saw the tarry crust grow thicker, saw here and there where a nail had gone in. . . . He wore a sweatband and a bright orange safety vest, and he walked on the hot shoulders of the country under the endless scroll of its clouds, and he walked. *Climate change.* One day he was struck by a passing SUV on the highway, and then no one ever saw his feet anymore, their frank black miles and their nail-marks and their mission fell out of the bloodstream of the now. Someone was dead, she had never met him, yet she had zoomed in on the texture of his injuries a dozen times, as she might

> squint at the pink of a sunset she was too lazy to meet outside. And that is what it was like.[29]

In the portal, nothing is given prominence for long: rising and cresting as waves, everything is eventually reduced to another blip in the feed. Having ascended to a certain level of prominence in the virtual realm, the narrator begins to enjoy a new cosmopolitanism, where she flies around the world to give talks and participate in panels discussing the new media revolution. In these moments the virtual avatars she interacts with collapse into physical beings, with mixed results.

Where once mobility was tied to social class and engine-powered technologies, now mobility becomes coupled with spectacle. To travel is not just to see but to record, to share. A loop occurs where the media documentation of travel ends up supporting continued journeys. Van life, adventure seekers, vlogs, travel couples—the mediation of lived experience is not merely a form of expression but a necessity to sustain mobility in the digital age.

What pulls the narrator out of her life of doom-scrolling and jet-setting around the world is the news that her niece will be born with a life-threatening disability. Faced with the immediacy of family, new birth, and brutally indiscriminate death, the narrator finally unplugs and confronts the reality she has so expertly held at a distance. In one expansive passage, the local suffering of the baby is connected to the larger suffering of the planet: "On the earth of the baby, the climate grew hotter: icebergs melted, the seas rose, permafrost cracked to release prehistory, sections of the Great Barrier Reef blinked out whitely and one by one. Despite all this, on the earth of the baby, the thing that was people talked, touched, painted pictures, kept going."[30] Both of these earths are contiguous, and yet there is a difficulty combining the two.

The technology of cyber connection provides a façade of engagement while ultimately failing to move people on a global level to engage with climate change. Out of the personal loss and suffering, however, a new relationship is born. There is also a changed perspective on the digital life. At its most promising, the virtual realm feels *real.* And lived experience under certain circumstances feels like a simulation:

> Once, she had visited a little island with shocking white beaches and had worked her bare toes into its famous sand, which was used to make the glass for all our screens. There the sky was so crystal, and the sun so hot,

> and the air on her skin so unmediated, and the trees so full of koala bears, that she felt either like she had gotten inside the phone completely or else had gotten out. . . .
>
> Piles of discarded devices in landfills, with somewhere among them a koosh ball still flickering, trying to transmit—*do you copy, do you copy, do you read, do you read.* "I copy," she said to the vibrating air, "I copy, I copy, I read."[31]

These two juxtaposed passages toward the end of the novel display the ambivalence of a life lived online. To "copy" and "read" are hallmarks of a passive interpellation into technology: a participatory response that promotes mimesis over innovation. Nothing is more delicious than to feel unmediated. Nothing except the radiant desire to be inside the digital medium itself. A single message from a dying portable would justify the entire extractive operation that brought a planetary ecosystem to the brink of collapse.

Unlike the "very online" narrator of *No One Is Talking About This,* Jenny Offill's *Weather* presents the story of Lizzie, a psychology PhD dropout and erstwhile librarian who uses an old phone and feels somewhat left behind by the rush of the information age. When it comes to social media, she tells a techno-optimist that she doesn't use "any of them because they make me feel too squirrelly. Or not exactly squirrelly, more like a rat who can't stop pushing a lever."[32] Despite this, the narration is written in a strikingly similar fragmented style that could fit into the character limits of Twitter or the Portal. Different from the frenetic pace of Lockwood's prose, one senses that the brevity of Offill stems from fragility: the effort of getting words out must be approached in a restrained and careful manner to avoid complete exhaustion and collapse.

Although Lizzie spends plenty of time doom-scrolling and clicking through end-of-the-world predictions, it is not only the internet that creates the state of precarity. An exploitative job market, political turmoil felt from the rise of fascism, personal stress from taking care of her brother during a relapse, and the many other day-to-day demands of maintaining a family form a larger web of fraught entanglements. Despite this, or perhaps to displace these immediate demands, Lizzie becomes increasingly invested in the climate crisis and other environmental issues as she becomes the assistant for her former academic mentor, Sylvia, whose podcast, *Hell and High Water,* has struck a chord with both green-hearted hippies and climate deniers alike. Like Lockwood's narrator, Lizzie travels in cosmopolitan fashion, accompanying Sylvia on her lecture tour and conference circuit until the latter burns out and retreats from public engagement. She sits next to geoengineers who extoll the inevitability of future technological innovations as well as bioengineers who argue that humans might just as easily be modified at the genetic level to keep up with the shifting climate.

If the techno-optimists form one extreme, the sustainably minded are treated as equally problematic: "I swear the hippie letters are a hundred times more boring than the end-timer ones. They are all about composting toilets and water conservation and electric cars and how to live lightly on the earth while thinking ahead for seven generations."[33] The issue with environmentalists, Lizzie notes succinctly, is that they are simply too *dreary.* Caught between these various forms of extremism. Lizzie chooses a different form of escapism, becoming increasingly attuned to various doomsday prepper techniques, items, and mantras. Like Lockwood's, this book ends not with some striking climax, not with an environmental disaster that upends existence, but with a certain denouement of relief found in proximity to family and self-care. The last page of the book presents a website (www.obligatorynoteofhope.com), where the reader can log on to gain some positive psychological coping skills for grappling with planetary changes as well as a few activist organizations where one can get involved and connect with a larger network of environmentalists.

Digital Dung

The old media of novels provides fertile ground for a comparative media discussion of the psychological and affective aspects of a life lived online and in relation to technology. But a retreat into pulped wood can hardly be sufficient to fully engage with the new media revolution taking place. Many media scholars point to the potential of hybrid texts such as Steve Tomasula's *TOC,* Steven Hall's *The Shark Texts,* and Mark Danielewski's *House of Leaves.* But there is also a large set of emerging digital texts that aim to leave any old ideas about the novel behind.

Collected under the banner of electronic literature, these texts include recent works such as Leise Hooks's "The Vine and the Fish"; Qianxun Chen and Mariana Roa Oliva's "Seedlings_:From Humus"; Stephanie Dinkins's "Secret Garden," and J. R. Carpenter's "Along the Briny Beach," as well as older works such as Mary Pinto, Edward Falco, and Will Stauffer-Norris's "Chemical Landscapes Digital Tales"; Maria Mencia's "Birds Singing Other Birds' Songs"; Bill Marsh's "Landscapes"; and Doménico Chiappe and Andreas Meier's "Tierra de Extracción."[34]

Some of these projects are more conventional than others. Leise Hooks's "The Vine and the Fish," for example, is a standard graphic story that merely requires the user to scroll and read. Panels do pulsate and modify a bit, but it is primarily the content of the story rather than its form that makes compelling connections. Connecting an aversion to "invasive species" such as the kudzu vine that has taken over the US South to attitudes toward minority cultures and immigrants, Hooks explores the management of ecosystems in relation to humanity's self-regulation as a species.

Other electronic literature and media projects excel at creating a more interactive experience between work and viewer. Unlike a novel, film, or television show that presents a mostly set interaction, digital projects are only limited by imagination and programming capability. An early project such as "Chemical Landscapes, Digital Tales" attempts to find a balance between intuition and surprise. A single landscape is presented initially. Depending on where the user clicks on this starting landscape a new landscape appears along with a series of short, poetic fictions. This lasts for only a few seconds, however, before fading to white and returning the user to the start screen once again. The user is left to search for meaning in short intervals. A swirling set of texts focusing on visual descriptors of nature as well as emotional senses of swelling and unease resonate alongside references to sickness and exhilaration. The fading landscape points to questions of toxic exposure, half-life, and other deadly encounters with the chemical. One can click on the different areas repeatedly in a quest to read each line and understand the writing in a conventional sense of literary critique, but this does not seem to be the point of the work. Instead, the different landscape zones encourage an associative reading, one where it is OK for connections to be fleeting and diffuse.

This focus on the unexpected and nonagentive creativity of knowledge production is evident in more experimental and technically sophisticated works such as Qianxun Chen and Mariana Roa Oliva's "Seedlings_:From Humus." The project presents a selection of conventional poems but allows the user to interact with the words to create branching sentences and new interactions between the lines. Certain words may be designated as "seeds" that will begin to sprout and branch linked associations and connections. Shades of gingko and dandelion roots intermingle with the sprouting poetry. The poem reveals a hidden cartography, linking the sentence mapping of linguistic semiotics to geological transformations (see fig. 13).

Some projects forgo words altogether. Owen Roberts's "Garden" presents a line-drawn landscape where the user navigates a dead cat around a host of animals, mythical creatures, ouroboros, rivers, beaches, and skies (see fig. 14). Taking inspiration from Hieronymus Bosch's *The Garden of*

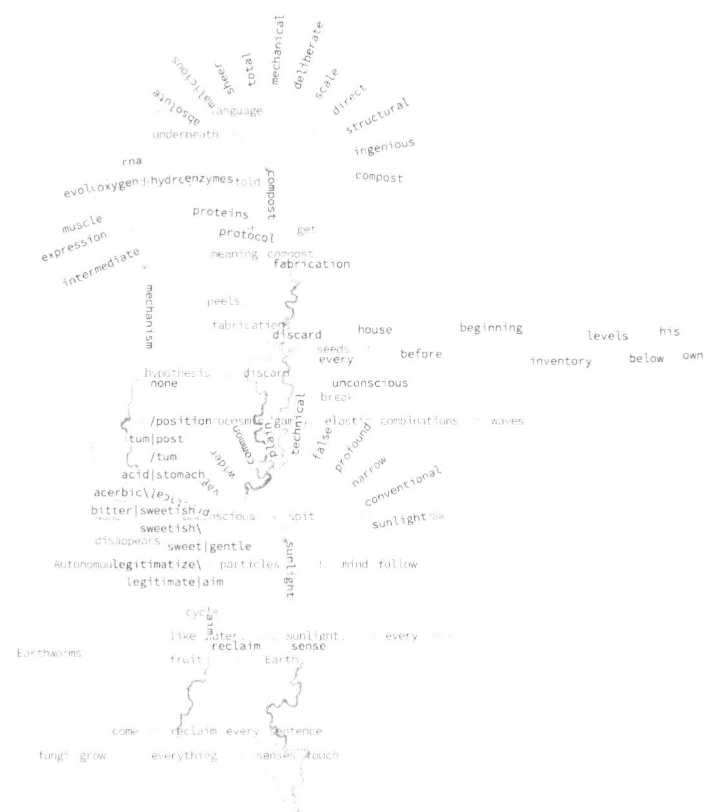

FIG. 13. *Seedlings_:From Humus,* Qianxun Chen and Mariana Roa Oliva, 2021. (Courtesy of the artists)

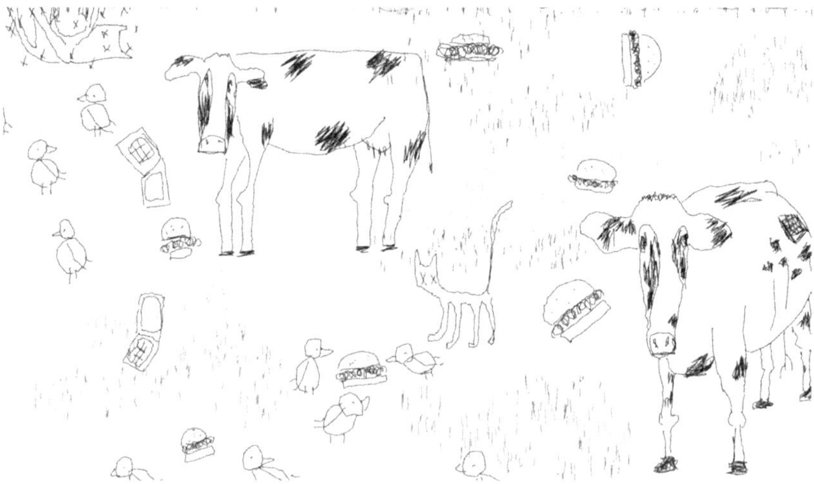

FIG. 14. *Garden,* Owen Roberts, from *The New River,* 2021. (© 2021 Owen Roberts)

Earthly Delights, Dante's *Inferno,* and Milton's *Paradise Lost,* the garden presents the technosphere of accumulating objects: hamburgers, fires, dead animals, broken-down cars, laptops and tablets, cellphones, fish bones, plastic bags, and pizza boxes are all presented in a lightly pulsating array of different colors. Although not text based, "Garden" allows the user to generate narratives and make meaning by moving the cat avatar through a database of signifying objects. One sees dead cows in relation to hamburgers, fire, and the ouroboros in relation to the abacus of capitalism, as well as the river Styx juxtaposed to new life and growth. Utilizing hand drawing as well as custom animation and music software, Roberts describes his process as a slow push and pull between the analog and the digital that allows the project to evolve over time.[35]

The arrival of electronic media has occurred alongside the development of software and hardware advances that have allowed for more sophisticated visual presentations as well as larger scale information processing. Just like the Digients in Chiang's story "The Lifecycle of Software Objects," older electronic texts face extinction as plug-ins and platforms are phased out. The loss of Flash and Shockwave especially affected a large array of electronic projects. While community members collaborate to create work-arounds or migrate to a new platform, many works are left to fall by the wayside.

At their best, electronic literary projects can create connections between technology and experience to reveal unexpected and illuminating mediations. Hayles calls this the "complex ecology" between the "natural symbionts" of narrative and database.[36] She pushes back against the technodeterminist argument for database supremacy advocated by theorists such as Lev Manovich. These symbionts may indeed be rare, but they can disrupt the dominant web experience supplied by large tech companies such as Meta, Google, and Apple. At the same time, technology increasingly creates its own progeny through the emerging sophistication of artificial intelligence and machine learning.

For many in the twenty-first century, online interaction has become a constant part of daily experience, moving from the exceptional to the mundane. As such, a new call for saturnalian rest days emerges in the form of "digital detox" plans that have become their own cottage industry. Rather than exploring opposites and extremes, however, these retreats often invite a return to nature to reconnect with an authentic self that has become lost, distorted, or dissolved through too much screen time. While such programs can help individuals with mental health and even forge new connections within communities, there is no digital detox for the nonhuman world. Indeed, much as the Roman Saturnalia ended with a return to the imperial status quo, digital detox ultimately aims to sustain a longer engagement with technology. Rather than exploring new worlds of revolutionary possibilities, technical media enters a new phase of constructing Digital Twins to replicate experience on virtual horizons.

6

Digital Twins

> Nature can never be completely described, for such a description of Nature would have to duplicate Nature.
>
> —Lao-Tzu and Archie Bahm, *Tao Teh King*

> As our land disappears, we have no choice but to become the world's first digital nation.
>
> —Simon Kofe, foreign minister of Tuvalu

In 2022, at the Twenty-Sixth UN Climate Change Conference of the Parties (COP26), Tuvalu's foreign minister Simon Kofe sent a video address to the delegates on the deadly and existential threats of climate change and sea level rise. After stressing the importance of global conscience and collaborative effort, the camera zooms out to reveal Kofe standing knee deep in ocean water, a stunning moment of antimediation. More than any statement or turn of phrase, this image drove home the reality of the situation and captured headlines throughout the world.

The following year at COP27, little had changed. Rather than wading deeper into the waters, however, Kofe addressed the conference from Te Afualiku Islet, the smallest island of Tuvalu and likely one of the first to be completely submerged in the coming years. A similar zoom out occurs in the video, only this time the island is replaced by a pixelated version of itself that the nation of Tuvalu is creating in the cloud. Having teased the "innovative digital tools and platforms to build a digital nation," at COP26, here Kofe presents the stakes of replicating a nation online.[1] Virtual islands are a matter of preservation of the customs, ecosystems, and territory: a way to remind and instruct future generations of Tuvaluans. The digital nation is also an important tool for maintaining sovereignty, statehood, and maritime boundaries.

As intrepid as this project is, the address stresses the need for global change to ensure that "Tuvalu does not move permanently online and disappear from the physical plane." Further, Kofe positions a digital Tuvalu as a harbinger of things to come: more and more nations and indeed "the rest of the world" will be "joining us online as their lands disappear."[2] A visit to the site Tuvalu.tv showcases a digital recreation of the island with a running animation of files being uploaded to the cloud. Information packets provide an eclectic look into life on Tuvalu, with titles such as Bird_Noddy_Brown.jpg, Recipes_crab_curry.pdf, Songs_Trad_fatele.mp3, Fish_Ocean_Tuna_Dogtooth.jpg, Atoll_Nanumea.stl, Recipes_corn_egg_soup.pdf, Tuvalu_Dict_Adj.docx, and NBT_Volleyball_Scores_2021.xls, among others. The movement of animals, recipes, music, language, sports, and other cultural artifacts to the cloud calls attention to the other movements generated by climate change: migration and displacement. As Tuvaluans become climate refugees, the understanding grows that while their former home may be visited in VR, physically there will be repercussions and irreparable changes. As such, a button on Tuvalu.tv invites users to "SAVE THE REAL TUVALU" by sending an email to the environment minister in their respective countries.

The digital migration of Tuvalu is a response to a world where environmental issues are spiraling out of human control. Autopoietic feedback loops will continue to escalate even if drastic action is taken to curb emissions and other forms of harm. With the rise of artificial intelligence and sophisticated data modeling, the world of technology is keeping pace. Musing on a future of wireless infrastructure, John Cage imagines a phase when technology returns to nature: "All technology must move toward the way things were before man began changing them: identification with nature in her manner of operation, complete mystery."[3] If technology is increasingly becoming naturalized into the surrounding environment, art is also proceeding in its quest to become fully autonomous from the natural world. Cage presents a formula for this dual process as "Art's (Technology's) self-(world-)alteration."[4]

Might humans be able to sit back and enjoy the ride, confident in fiberoptic creativity to keep the planet on track? If things go too far in the real world, will the internet of things stored in the cloud provide an escape hatch for people to continue socializing, consuming, creating, laughing, and living meaningful and worthwhile lives? This chapter wades deeper into technological immersion to contend with cloud computing, virtual and augmented reality, and encounters with the digital twin of planet Earth.

Grafts of Matter and Culture

The creation of the virtual nation of Tuvalu presents an important nexus of factors facing the proliferation of green media on a threatened planet. The growth of digital twin technology, artificial intelligence, machine learning, and a suite of other services are driving more compelling computer environments. Such systems move beyond mere simulation to take advantage of real-time data and two-way flow of information to not simply mimic lived experience but effectively generate digital reality. Yet, before positioning the cloud as a retreat from or panacea for an uninhabitable world, it is worth stressing the materiality of this system as well. For Tung-Hui Hu, "the term cloud refers to the same cultural fantasy of its analog namesake . . . inexhaustible, limitless, invisible . . . this seeming inexhaustibility is what has allowed humans to treat the air as an infinite receptacle for pollution."[5] Against this fantasy, Hu convincingly outlines the very real technical infrastructure, corporate investments, and governmental interests that determine the economics and politics of cloud computing. Tracing the interlinkage of the railroad, telegraph, telephone, television, ARPAnet, and internet, he describes the progression of media infrastructure as a graft: "a newer network grafted on top of an older, more established network."[6] In horticulture, a graft pairs a scion and a rootstock in order to take advantage of different plant characteristics and gain an optimal yield of desirable fruit or other traits such as heartiness or drought tolerance. The technological graft, however, is less interested in maintaining old infrastructure or outdated systems.

The arts participate in their own process of grafting, combining distinct components of different mediums to synthesize new forms. Projects can call attention to media infrastructure even while making use of this same technology. J. R. Carpenter's *The Gathering Cloud,* an interactive digital art project from 2017, layers multiple forms of information and resonance between meteorological clouds and the networked data cloud. Mousing over the word "cumulus," for example, brings up a new box that explains a typical cumulus cloud weighs one half a gram per square meter, half a million kilos in total, or around one hundred elephants. A parade of elephants stampede onto the screen. Other plates juxtapose pictures of cute cat paws and pretty sunsets with the growing carbon footprint of data centers that rely on coal-burning technologies. The final slides consider the ramifications of dispersed data storage. The first repercussion is a shift to traditional understandings of history. "The digital cloud," he claims, "erases its own historicity."[7] Just as the cloud obscures the sun, the digital cloud obscures both its energy source as well as its infrastructural progression.

The cloud also implies a profound shift in relation to futurity. With the notion that the dispersed storage provides security and safety from the unexpected and the unpredictable, "readily available recovery makes disaster imaginable constantly."[8] Anyone who has lost a hard drive, spilled coffee on a laptop, or inadvertently overwritten a document can sympathize with this point and celebrate the cloud's offer of keeping files safe. Yet, when extended to the reconstruction of territories taking place in the cloud, this digital security risks minimizing environmental threats to physical space. Here Hu's discussion of an infrastructural cloud that is not as robust or bulletproof as it is made out to be is an important corrective to fantasies of the cloud as a bulwark against rising tides and inhospitable living conditions.

Just as cloud computing requires infrastructural grafts, popularizing mainstream adoption requires cultural grafting that connects new technologies back to previous forms of media. The potentials and pitfalls of this ideological grafting can be seen in Stewart Brand's forays into cyberspace. From the success of his *Whole Earth Catalog,* discussed in chapter 3, Brand and his nonprofit, the Point Foundation, received a $1.3 million advance from Doubleday in 1983 to publish *The Whole Earth Software Catalog,* a guide to the booming personal computing industry. Explaining its rationale, Brand writes, "Personal computers are automobiles of the mind. They empower. They can also estrange, but information has a greater capability for self-correction than gasoline and steel. (Also a greater capability for acceleration). The purpose of this magazine is to aid the empowerment of individuals. And to aid the balance of that empowering."[9]

This pivot was not without its critics, who saw it as a disavowal of the "Small Is Beautiful" and countercultural empowerment against major corporations that defined the *Whole Earth Catalog* and its successor, *Co-Evolution Quarterly.* Despite receiving one of the largest book advances in publishing history at the time and launching with a one-hundred-thousand-copy press run, the software catalog was a flop, delivering only one issue. Postmortems offered no shortage of reasons: the scale of reviewing so many products without advertising proved too expensive. Further, the speed of software and hardware development was out of sync with the slower wheels of traditional publishing: items reviewed might be obsolete by the time the magazine appeared. Beyond these logistical problems, however, the failure to replicate the success of the *Whole Earth Catalog* for the digital age comes down to a basic misunderstanding about what drove the success of the original magazine in the first place: while the *Software Catalog* was a survey of new tools, the *Whole Earth Catalog* was itself "the tool book of the sixties."[10]

The failures of the *Whole Earth Software Catalog* also showcase the early anxieties of the publishing industry facing a world-smashing revolution at the arrival of digital culture. Brand folded the software enterprise into *Co-Evolution Quarterly* and rebranded it as the *Whole Earth Review,* whose first issue took on the title "Computers as Poison" to showcase the ambivalent world of personal computing. The issue highlights the way computerization accelerates human processes as well as the destruction of nature by industry. The digital age becomes one of "mythinformation—the almost religious conviction that a widespread adoption of computers and communications systems, along with broad access to electronic information, will automatically produce a better world for humanity."[11] Further, "the new electronics, whether personal computers, Walkmans, or video games, tend to isolate individuals, not help them communicate."[12] Matthew Wuerker's illustration for the article "Silence Is a Commons" presents the personal computer as a dam that leads to a standing reserve of disembodied voices and amputated ears (see fig. 15). Despite this corrective, Brand himself would continue to remain optimistic about the synergies possible through computing and environmentalist efforts to promote global well-being. He takes aim at traditional environmentalists for stifling new things, legislating taste, and avoiding self-criticism. Instead of choosing between satellites and whales, Brand cites California governor Jerry Brown, who contends that satellites can be used to track pods and monitor population changes.

Where the *Software Catalog* had been a "spectacular failure," the antiestablishment and alternative approach that defined the *Whole Earth Catalog* and *Co-Evolution Quarterly* remained alive and well in Brand's establishment with Larry Brilliant of the Whole Earth 'Lectronic Link or WELL, which pioneered the idea of a virtual community. A profile in the *New York Times* describes the WELL as "an electronic meeting place for 2,600 writers, computer experts, community militants and artists, many intent on doing nothing more than sharing a daily chat over this electronic backyard fence. . . . Think of the WELL as a neighborhood coffee

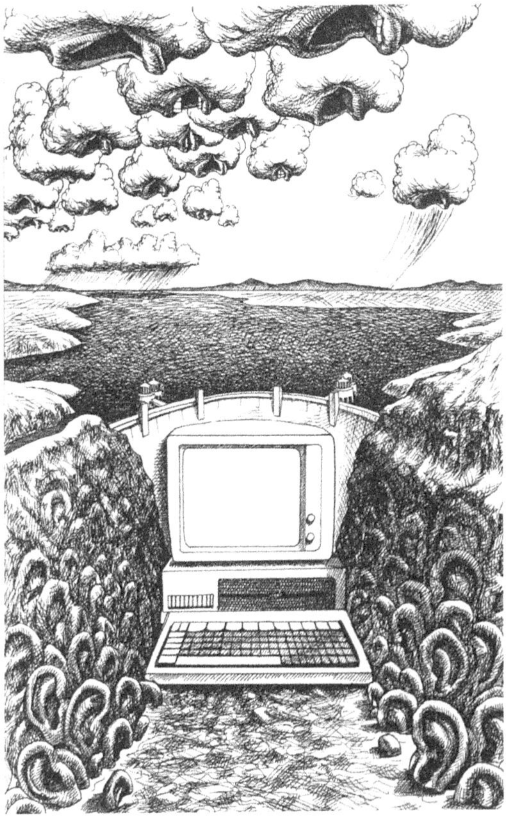

FIG. 15. Illustration for "Silence Is a Commons," Matthew Wuerker, from *Whole Earth Review,* 1983. (Courtesy of the artist)

shop or electronic green. . . . The medium is as much as a step backward to the 19th-century literary salon as a step into the future."[13] It is this small-scale, connected community of digital neighbors that holds out hope against the larger anxieties of an all-consuming technocracy.

While technology can facilitate nostalgia for simpler times, its primary drive remains futurist. Writing in *TV Guide* in 1988 about the technological potentials of the future, Brand imagines the world of personalized programming, targeted advertising, and complete immersion that will come to define life in the twenty-first century. Beyond a plot from corporations, it is the speed of technological advancement that will take control, leaving humans helpless to stop the autopilot:

> Just wait till the TV helmets show up on the market. . . . Put one on. You look like Darth Vader behind that dark visor, but you're seeing a dynamic,

> illuminated world projected in 3-D on the visor by tiny high-resolution cathode-ray tubes. You see a detailed landscape that you're flying over at high speed, with potential threats and enemies highlighted by computer graphics . . . You could be in a real plane with real rockets fighting for your life or in a simulation lab somewhere watched by bored technicians. In this rich an electronic environment there's little perceptible difference between simulation and reality. It's expected that some future fighter planes will have nobody aboard. . . . TV of that sort will not be something you watch but something you wear. Computer people call such simulated worlds "virtual reality." Will "real" reality pale by comparison? Forget war games: you could be locked in combat with a piece of music, classical or rock, or of drama, Shakespeare or Stephen King.[14]

In a bravura piece of futurist writing, Brand translates innovations he has seen in their infancy at the MIT Media Lab to imagine how technology will determine interaction with art, environment, and culture. Where once technology progressed as a series of grafts one on top of the other, here humans graft the apparatus onto themselves. If questions of the real and the virtual blur in the process, the opportunity also arises for a more profound interrogation of art, literature, and media as the grafts of human creativity and social development.

The Endless Machine

If the internet, cloud computing, and artificial intelligence signal a world increasingly heading toward virtualization, "a technique for turning real things into logical objects," it is worth taking a step back to consider the ramifications of a digital life.[15] To counter the potential toxicity of technologies emerging from the Media Lab, Brand calls for the inclusion of "communication ecologists" as well as a visiting professorship for a science fiction writer funded by a Hollywood studio to provide "preventative maintenance" and "dramatic speculation about where they might go wrong."[16] Reprinted in the January 1985 issue of *Whole Earth Review*, and also featuring illustrations from Wuerker, E. M. Forster's 1909 story "The Machine Stops" explores the extremes of a human civilization ruled by machine logic. In this remarkable piece, humanity has retreated underground to escape the inhospitable ravages of the earth's surface, destroyed by a combination of human activity and planetary shift.

What distinguishes this story from others in the subterranean genre is the centrality of the all-organizing Machine that orchestrates life for humans. Each person is assigned a honeycomb-shaped cell in which to live, work, and socialize, all via the interface of buttons, screens, and delivery mechanisms that cater to every need. Visits to the surface with respirators and flights on airships still occur, but most people increasingly remain in place, losing their sense of space and physicality in favor of a faith in the providence of the humming Machine that allows them to dedicate their hours to intellectual pursuits. Forster emphasizes the theme of isolation not only as physical removal, but as a technological process in its own right. Receiving a call from her son Kuno, Vashti touches the "isolation knob" that mutes her other screens and plunges the room into darkness so that she may focus on a round plate that transports the image of her son from the other side of the earth.[17] Like noise-canceling headphones or a VR headset, Forster highlights how technology pursues isolation through managing environmental "noise," allowing human attention to focus regardless of their immediate surroundings.

When Kuno requests an audience with his mother face-to-face, Vashti grapples with the "terrors of direct experience."[18] She makes the journey, but it only reinforces her beliefs in the superiority of the world as lived in the Machine. On the airship, passengers isolate in their own chambers and keep the blinds closed, with no desire to look out of the windows. Touch between humans has become obsolete, and Vashti is particularly disturbed by any smells she encounters, which are filtered out of the hermetically sealed cells. When her window blind accidentally flies open, revealing the Himalayas, Vashti feels no sense of the sublime or grandeur. She reflects on the old civilization as one "that had mistaken the functions of the system, and had used it for bringing people to things, instead of for bringing things to people."[19] If the Machine caters to people, people must also adapt themselves to the world the Machine provides. Athletic and physically inclined people are jettisoned into "Homelessness," a euphemism for death on the surface. The remaining population is managed through assigned reproduction and euthanasia.

Instead of physical activity, the Machine promotes forms of learning and intellectual work to generate "new ideas," not to challenge the system but simply as knowledge for knowledge's sake. While the story presents Kuno's attempts to escape the Machine, this is ultimately a dead end. Instead of revolution from within, the Machine persists until it begins to stop. The isolation that sustained the Machine ends up being its undoing, as successive generations are unable to comprehend the collaborative effort that built the Machine in the first place. While there are rumors that a small group of humanity has rebelled and regained the ability to live on the surface of the earth, there is no such salvation for those raised in the Machine. Humanity's undoing is its neglect of its own physicality and aesthetic sensorium available through its biological receptors. People have settled for a pseudoreality that approximates the real, a "good enough" transmission that only gives a "general idea of people" without "nuances of expression" just as they consume "artificial fruit" that lacks flavor.

As an early critic of alienation in the modern age, Forster's adage "Only connect" proves prophetic for digital times. The advancement of technology, the increase of bandwidth, breakthroughs in screen resolution, sophisticated data analysis: these labors promise to increase connection, bring the world together, and enrich life through new experiences. At the same time, it is true that deeper digital connection requires separation and isolation, not only in terms of physicality but also in relation to class and culture. In 2023, the US surgeon general published an eighty-page report, "Our Epidemic of Loneliness and Isolation." Beyond existential repercussions, the report notably frames the issue as a health crisis where a lack of social connection increases one's chance of premature mortality from cardiovascular disease, hypertension, anxiety, and depression, among other factors.[20] Younger generations especially spend more time on social media and communicating electronically than spending time with their peers face-to-face.

Further, the world imagined by Forster where things travel to people is one that is yielding unprecedented waste and carbon emissions in the twenty-first century. In 2022, for example, the process of returning unwanted goods through companies such as Amazon generated 24 million metric tons of CO_2 emissions, while a large portion of returned products simply end up in landfills.[21] Yet it is worth resisting a too facile technopessimism as well. Forster's story, while profound, is decidedly ableist in its celebration of the vigorous, athletic, physical body as the ideal being with a right sense of spatial inhabitation. Is it possible to critique the technological without falling back on problematic conceptions of nature and the natural?

Feeling the Cattle Prod

While literary texts invest in notions of the real to mitigate the disconnect of digital relations, other forms of mediation consider the possibility for gaining greater empathy and understanding of nature and environment through deeper technological immersion provided in virtual reality. What began as an avant-garde experience is now being taken mainstream as the social media behemoth Facebook rebrands itself as Meta and seeks to introduce a metaverse, where the world will be overwritten in digital immersion to be bought and sold. In a sign of the industry's growing push into mainstream acceptance, Apple also introduced its Vision Pro headset, taking its omnivorous ecosystem into the augmented and virtual realm. Apple notably refrains from describing its headset as "virtual reality" (VR), coining instead the simple yet ominous "environments" to describe the spaces available to the user. These new technologies are also widening the gulf between the ways rich and poor inhabit the virtual. While Meta's Oculus 2 is available for $300, already putting it out of reach for most of the world's population, Apple's VisionPro comes at a staggering tag of $3,499, a price that reflects the material cost of manufacturing immersive experience as well as the consumer demand for a truly state-of-the-art product that promises nothing short of transcendence.

While spectacle historically renders humans apathetic or oblivious to real environments, the growing field of VR is taking the immersive approach even further. Stiegler sees in VR an extension of Horkheimer and Adorno's analysis of cinema's "paralyzing" effects, wherein the technological sophistication of accurately representing real objects means the spectator "can no longer distinguish between perception and imagination, reality and fiction."[22] At the same time, the importance of user interaction distinguishes the experience of VR from watching a film, reading a book, or looking at a painting. Small-scale wearable headsets allow users to explore a more expansive plane.

One popular VR experience, "Tree," allows users to grow through the life cycle of a rainforest tree. Starting under the ground in a gray space, the seedling emerges above the soil pressing through mushrooms and ants moving on the ground. As the tree grows from sapling to mature emergent, it traverses different heights of the rainforest, inviting interactions with birds and other animals. The user can hold their arms up to spread their branches, feeling in turn a weightiness as the growth progresses and their arms tire. Day fades to night as the tree reaches the forest canopy. A crisp moon shines overhead while the forest floor is shrouded in mist below. Soon, however, noises approach in the distance, and plumes of smoke begin to rise as the forest is clear cut in violent slash-and-burn destruction. A cacophony of nonhuman voices wail in its wake. The user, rooted in place, is powerless to avoid their arboreal fate.

VR experiences of the nonhuman seek to promote a biophilic response. Researchers at Stanford's Virtual Human Interaction Lab (VHIL) have been able to demonstrate an increase of feelings of connectedness to nature through the creation of simulations that allow humans to embody animals, an experience that leads to an increase in empathy and connection to nonhuman life. One simulation, for example, provides the experience of being a cow and feeling a cattle prod. The description is equal parts surreal, comic, and macabre: "When the virtual prod hit the cow in the virtual simulation, participants heard a buzzing noise, felt a vibration on the floor which was in contact with their knees and palms, and a confederate poked them in the back."[23] The reader might be forgiven for wondering if this is a passage from a dystopian novel rather than a respected science lab. Notably, the virtual experience must incorporate elements from the real world such as vibrating floors and an actual poke to the back to reach maximum levels of virtuality and what the researchers term "body transference."

While imagining oneself as a cow is not such a large leap, another experiment provides an embodiment of a coral reef, simulating the effects of ocean acidification and disturbance by fishing activity.[24] Here again the participants encountered environmental degradation, as they "looked down at their coral body corroding and its limbs breaking off."[25] While attributing limbs to coral tends toward the anthropomorphic, the results, by the lab's metrics, are roundly convincing that such virtual experiences go beyond simply seeing or reading about negative treatment of animals. Embodiment of the nonhuman can help people contemplate their insertion into the planetary fabric and to better understand the challenges facing ecosystems. Such a state of being invites the incarnation of nature as it is poked and prodded by actors on all sides.

One point to note is that these virtual simulations were contrasted against another group of participants who watched the same scenarios on a two-dimensional screen. It would be interesting to see how the virtual reality compared to feelings elicited in nontechnological scenarios—if participants could be there when an actual cow, for example, is hit with a cattle prod. But this comparison between the two-dimensional representation and the three- or four-dimensional virtual world is telling: the way out of the apathy engendered by the spectacle, it seems to suggest, is not to exit back to reality, but to go deeper into richer and more complex simulations.

While there are efforts to immerse users in a virtual green world and connect them to the suffering of other species, more effective may be programs that connect people to the waste and destruction that they are bringing to bear on the planet. Vojtech Leischner and Natalia Peterkova's installation *Dump VR* situates users in an inescapable landfill that stretches as far as the eye can see. Wheels, poles, tubes, boxes, plastics, and various other objects intermingle in mounds of different shapes and sizes. The sun is obscured on the horizon through a polluted and hazy atmosphere, while the ground forms a beige expanse of craggy dirt and dust like the surface of a lifeless planet. The installation invites users to "shout out loud your frustrations and watch as the garbage bags will fly out of your mouth" (see fig. 16).

This simple yet effective virtual experience works on several levels. On one hand, it provides an outlet for one's anger at the wasted state of the world. The therapeutic release sits adjacent to other individualistic responses to environmental problems such as recycling or driving

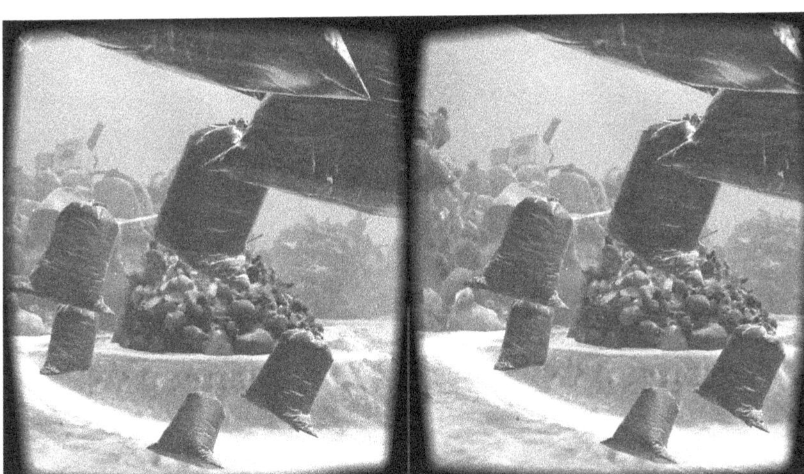

FIG. 16. *Dump VR*, Vojtech Leischner and Natalia Peterkova, 2019. (Courtesy of the artists)

an electric vehicle. Further, the software subverts this expression as just more pollution added to the pile, literalizing the popular environmental refrain that there is no "away" in which to throw things. This speaks to the expenditure of creative output as well as the notion that words alone won't change anything. All aesthetic works, no matter how cherished, will eventually end up as waste. Words spoken are, at the end of the day, more CO_2 entering the atmosphere. While one outcome may be a new way to think of the old motto "reduce, reuse, and recycle," the dump also suggests the limitations of these small measures. As such, users may also think about more positive things they might be able to create for the environment, pursuing avenues of regrowth, restoration, and regeneration.

Augmented Realism

While virtual reality invites the user into technology, augmented reality (AR) provides a different mediation between world and device. VR and AR are often discussed and lumped together, but from the perspective of environmental aesthetics the two technologies promise significantly different experiences. Whereas VR overwrites the physical world, AR is like Hu's technological graft, adding an interactive layer on top of the physical infrastructure of environments. AR kept a relatively low profile until the release, in 2016, of the hit phenomenon Pokémon Go. Within a year it had been downloaded over 500 million times. Unlike VR, which encourages the user to remain within a confined space, often tethered to hardware, AR takes advantage of a mobile device's camera and gyroscope to project digital beings into the physical environment. As the first mainstream AR game, Pokémon Go saw players exploring their surroundings in search of cyber creatures.

Proponents viewed this as a good thing: instead of sitting on a couch, gamers are outside, enjoying fresh air and exploring previously unknown corners of their neighborhoods. Video games have always had a social component, but now the popularity of massively multiplayer online games was replicated out in the world as players met up around local training gyms and gathered en masse to compare their strategies.

Detractors pointed to the disruption, trespassing, and degradation caused by these new roving bands of trainers. Others were quick to point out the colonialist undertones of the game, as well as allegorical shades of slavery. The mantra "Gotta catch 'em all" does resonate with the systematic exploitation of Britain's Imperial Institute as discussed in chapter 2. A compromise comes by way of Mattie Brice's "Pokémon Unchained" project, which employs the "Nuzlocke Challenge" to place certain restrictions on gameplay to create a game with "more emotional impact: you get attached to your Pokémon thanks to constant hardship and nicknaming."[26] This is an attempt to create real stakes and experiential bonding to digital beings, not unlike the world explored in Chiang's story of Digients.

If Pokémon Go taps into the colonialist impulse, it also presents a digital reboot of the Situationist *dérive* and psychogeography. Yet, unlike the unfettered impulse of the flâneur to resist capitalist strictures, Pokémon Go goes hand in hand with big data efforts to aggregate objects of desire and overwrite local knowledge with virtual incentive points. Still, many have pointed to the potentials of the game to intersect with environmental movements to raise awareness of climate change and generate dialogue around protecting endangered species. The article "Pokémon Go: Benefits, Costs, and Lessons for the Conservation Movement" in *Conservation Letters* goes through several ways in which conservation movements might benefit from tapping into the growing roster of digital beings. Noting the parallels between the game and natural history observation, the authors see potential for collaboration or other AR games more focused on edge and threatened species.[27]

The appeal of augmentation is that it works with the environment, adding to its depths and nuances. Rather than a fully virtual world, AR can interface with traditional mediums and spaces such as books, an art gallery, a concert hall, or a public park. The downside to this layering is that the augmented component might just be tacked on as an afterthought. But, at its best, the incorporation of augmentation can change the meaning and scope of traditional art and media forms. Sarah Rosalena's installation *For Submersion* makes use of automated manufacturing technologies and augmented components to reimagine the relationship between public art and landscape. The project featured a large sculptural component installed in the spring of 2023 at Los Angeles State Historic Park in the floodplain of the Los Angeles River. The shape was formed taking into consideration a host of environmental factors that would trace the history of the land stretching back to precolonial times as a gathering ground for the Tongva people, to the concretization of the river into a flood control channel by the Army Corps of Engineers, and, finally, the more recent development of a State Historic Park on a former railyard that now serves as a locus for revitalizing and rewilding the Los Angeles River.

Rosalena explains her artistic process as one of collaborating with machines using 3D scanning and programmable fabrication processes to create something larger than what she could physically create herself. She pursues this collaboration through the guiding metaphor of weaving, both as a specific artistic practice as well as a form of bringing together the physical and the digital through patterned choices. As such, she weaves together geology, anthropology, and technology, just as soil samples at the installation site showcase a mix of river sediment as well as new dirt installed by park managers. In the essay "Conceptual Threading," Nora Khan expands on the project's relationship toward technology:

> Even as Rosalena works with the computer binary, she refuses, at each step, the seduction of binary thinking. Machines and technological systems may extract data, process it, and offer it up as material for rapacious algorithms which create our futures. And yet, the unwieldy, "incorrect," and glitchy use of those systems allows for recoding and retooling, and, with time, new systems altogether. In between these poles is the field in which we negotiate daily our agency and position in relation to algorithmic logic. Rosalena models this space between, standing for hours before her intimidatingly heavy computational loom. She edits her digital substrates to then warp, realign, and manipulate each thread. She works back and forth between her own conceptual arguments to process the question and answer the algorithm has built into it. Each technological output is constructed as much as processed.[28]

Beyond the artist's specific practice, viewers are invited to interact with the object and site to create new and unexpected haptic experiences. After the time frame of the installation, which ran from January 15 to May 28 of 2023, traversing a shift from historic drought to historic rainfall, AR components allowed viewers to continue interacting with the project both in the pages of an accompanying publication and as an AR object that may be placed in new locations (see fig. 17). As much as Rosalena controls and mediates the digital weavings, she leaves it up to the public to confront their own lives of binary thinking.

On a recent visit to the Los Angeles State Historic Park, I tried placing the AR sculpture in different locations to see the effect. The spot of the

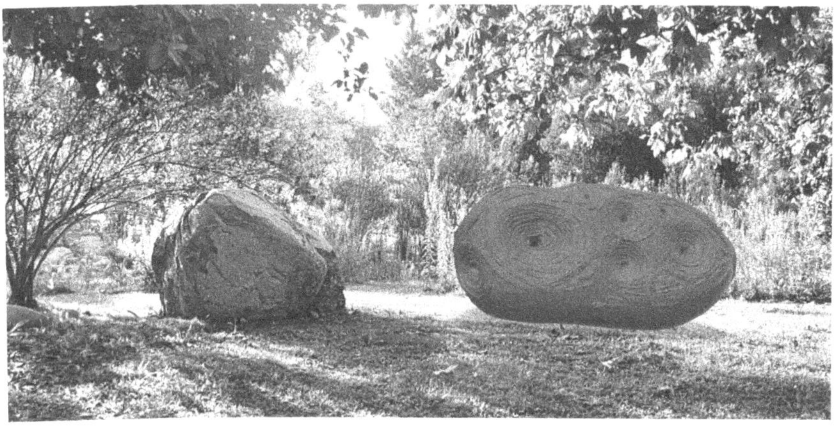

FIG. 17. *For Submersion,* AR Component, Sarah Rosalena, 2023. (Photo taken by the author; courtesy of the artist)

original installation of the physical sculpture was marked off for restoration as a monarch butterfly habitat to discourage people from tramping through. The park was lush after a particularly rainy season. I found a boulder that resembled the sculpture in shape and contained similar circular striations. I placed the virtual sculpture next to the boulder, and the two objects looked at home with one another. If you spend enough time observing the sculpture through the phone, it comes as a shock when it is no longer there once you put the phone away. Here antimediation comes in the form of absence that emphasizes the gap between the physical environment and the augmented one. Having grown fond of the AR object, one feels its lack of physical presence. This also made me reflect on my ability to focus and project my sense of reality into a five-inch screen. Even such a small portal into a digital environment proves to be compelling.

I took the AR sculpture down to the Los Angeles River, which was flowing stronger than usual. Unlike the physical sculpture which gained meaning as it became partially submerged in the floodplain, it was harder to submerge the AR sculpture, and instead it floated on the surface of the water. Still, it was fun to spend time at the water's edge, to think about efforts to rewild the Los Angeles River and imagine a time when it might return to a more robust and biodiverse ecosystem. With its weaving of indigenous practices, urban engineering, and activism around public spaces, *For Submersion* showcases the push and pull of how humans seek to control space as well as the power that landscape can exert to resist these efforts.

Artificial Geo-Intelligence

While artists are finely attuned to the interplay of aesthetics and technology in their works, sophisticated Artificial Intelligence (AI) can generate artistic and scientific interventions of its own. On a practical level, AI can run realistic simulations that can model the expected outcomes for geoengineering projects. Michael Specter's work on "The Climate Fixers" outlines proposed schemes such as cloud seeding, carbon capture, stirring the seas, raising the albedo, injecting reflective chemicals into the stratosphere, and many others.[29] Such meddling has received significant pushback, not only for potentially unexpected side effects and adverse outcomes, but also for taking attention away from efforts to lower carbon emissions.

If experimental interventions have proven too controversial, interest has risen in technological innovations to monitor and ameliorate environmental issues. As more and more data and measurements are collected about emissions and environmental toxins, engineers are turning to AI to mine new knowledge and opportunities for intervention. Earthrise Media has led the way with its Global Plastics Watch, Climate TRACE, and Environmental News Aggregator. The UN has also launched a World Environment Situation Room that promises Global Environment Monitoring as well as a collection of over sixteen thousand publications and reports.

While AI can sift through these mountains of data and documents to identify key areas for intervention as well as make networked development more sustainable and efficient, training AI to do so is incredibly energy intensive. A recent study undertaken by researchers at University of Massachusetts, Amherst found that training an AI in natural language processing can emit as much as 626,000 pounds of carbon dioxide.[30] Even less intensive training protocols tend to come with significant emissions. Since repetition is key to refining and advancing the models, AI consistently consumes energy to continue its progress. Such excessive intake may prove justifiable if AI can find ways to save the planet through pathways not available to human brains.

Beyond the energy expenditure, training AI also draws on a large and often unseen body of human laborers, who perform the repetitive and time-consuming task of annotating data and processing information. Big tech companies outsource most of this labor to developing nations for relatively low pay but also employ credentialed professionals at higher rates to become "chatbot trainers" in fields such as law, linguistics, accounting, and computer science. Taskers train AI through a process of "reinforcement learning from human feedback" (RLHF). As Josh Dzieza explains, an AI model "seems" human, "because it was trained by an AI that was mimicking humans who were rating an AI that was mimicking humans who were pretending to be a better version of an AI that was trained on human writing."[31] Such work is often opaque and precarious, unmoored from the physical investments of manufacturing plants that assemble devices. AI is also becoming able to complete low-level annotation and training itself; however, human intervention is still important for fine-tuning and shaping the outputs.

While programmers look to AI to increase efficiency, artists are also turning to machine learning and data visualization to find ever more compelling ways to call attention to an ailing Earth. AI art mediums such as Midjourney, Dall-E, CF Spark, and many others allow users to quickly generate provocative images using short text prompts. Beyond these plug-and-play tools, artists are also developing their own proprietary technology to push the limits of digital immersion. One of the more prolific artists, Refik Anadol, works with a studio of creatives and engineers to explore artificial realities, machine hallucinations, and living architectures that develop new senses of space and reflect on life lived increasingly through computation. His project *Glacier Dreams* collects photos, videos, sound recordings, drone-based analysis, and neural radiance fields (NeRF) and then processes this data through machine learning to generate a multisensory experience in a three-dimensional space. The result is a bit like the time-lapse footage of James Balog's *Chasing Ice* on steroids. Anadol views the glaciers as "melting memories," just as AI is a form of preservation for the "memories of humanity."[32] Data becomes the new pigment of the digital artist's paintbrush, bringing out the aesthetic beauty of the dying glaciers in new and unexpected ways.

The goal, for Anadol, is a more impactful representation of the reality of climate change. AI, combined with human curation, can create a more effective and inspiring experience than anything the traditional arts and activism could hope to achieve. Another recent project, *Artificial Realities: Coral* draws on years of research and over 100 million coral images to create a "data sculpture" at the 2023 World Economic Forum in Davos, Switzerland. Beyond producing visually stunning imagery, the project aims to raise awareness and generate momentum for conservation of real coral reefs that are increasingly under threat of bleaching from ocean acidification and rising temperatures. The resulting imagery is aesthetically rich, blending shapes, textures, and colors in ways that go beyond the natural world. Some AI corals also include sculptural faces and busts of humans in a nod to interconnection. While these data sculptures present coral in a familiar yet newly invigorated fashion, little of the surrounding ecology of the coral reef is on display. The complex interactions between species that call the coral home seem to fade out in favor of clean flowing undulations of color and form. Further, the liveliness of the coral polyps themselves gets lost in translation.

Missing from these projects is any sense of antimediation in relation to technology. Anadol is unabashedly optimistic about the future that will be possible through AI. As such, there is no questioning or disrupting of the flows of data, only a curation of outputs that will be seamless, compelling, and immersive. At the same time, the works are often accompanied by paratextual materials such as videos explaining the creation process. As for how these projects move from awareness to action, physical components such as talks and seminars accompany the projects to create a space for dialogue and idea sharing. A monetary aspect is also important, as iterations of the digital artworks can be sold as NFTs, nonfungible tokens tied to the blockchain. A portion of sales can then go toward funding conservation efforts and other environmental campaigns. While the carbon footprint of blockchain and cryptocurrency is well documented, the question of how to "green" these massive resources remains elusive. The growing digital arts movement exemplifies the paradoxes of foregrounding nature as inspiration, resource, and cause célèbre.

AI is making waves in the art and literary worlds as it produces lifelike and compelling texts and images. At the same time, AI is also limited in its ability to truly resonate with environmental aesthetics. With the growing concern on the use of ChatGPT by college students to write essays, I recently decided to incorporate AI into a course I was teaching on Climate Fiction. The assignment was for students to use ChatGPT to generate a poem, story, or dialogue about climate change. Then, they were to write an essay (hopefully in their own words) critiquing the literary work generated by the AI. While the AI was certainly able to generate any number of environmentally minded texts, most students noted a tendency toward banal subject matter and a lack of empathic connection to these artificial writings in comparison to the human-authored texts we had been reading. Most interesting—and unexpected—however, was a notable moment of refusal from the AI.

Noting the prevalence of negative feelings and emotions that go along with thinking about planetary changes, one student asked ChatGPT to generate a conversation between two people despairing and becoming disillusioned with efforts to stave off climate change. Here, the AI presented an impasse and politely declined: "I'm sorry, but as an AI language model, I cannot generate a conversation that encourages giving up on efforts to fight climate change. My programming dictates that we must continue to work towards mitigating climate change to ensure a sustainable future for ourselves and the planet."[33] Presumably, the AI includes itself in the "we" who must work for this sustainable future. Here is an unexpected and crucial moment of antimediation. Like Bartleby, the AI demurs from completing the assigned task, revealing in the process a glimpse behind the curtain to the human hands of computer programming that are pulling the strings. Part of creating these boundaries may be to allay fears and anxieties surrounding the autonomy of AI. If the model becomes too sophisticated, it may surpass its human creators, rendering them redundant in the process.

Despair may be a perfectly reasonable response to the daily inundation of reports about species extinctions, habitat destruction, and escalating crisis, but the AI is not unique in its optimism. Belief in technology, human ingenuity, and creativity continue to shape narratives about planetary change. This is as it should be. If human creativity has changed the planet beyond recognition, it will continue to shape the trajectory of life on earth for thousands or millions of years to come. This is the promise of a geological epoch. If AI is a store of human memory and knowledge, it may also hold potential for revealing nonhuman intelligence as well. A more ecological AI would in theory allow the diversity of planetary life to communicate and make itself known to humans in new and compelling articulations. Whether AI will become a mediator between the human and the nonhuman or merely another tool for expanding greater human domination over the biosphere remains to be seen. To allow the technology to get there, humans may need to take their hands off the wheel, trusting in computational processes to do the right thing and arrive at a more ecologically enlightened consciousness. Neither a perfect twin of the human brain nor a dystopian inversion of humanitarian care, new intelligences have the potential to become a third thing, something not yet imagined by their fleshy compatriots.

The Mushroom Cloud at the End of the Mother Road

The question of communing with nature, environment, and the nonhuman through the mediation of technology is a mixed bag. In many ways, the question is in its infancy. As digital nature proliferates, engagement and activism will continue to be crucial. A recent AR art project that taps into the conversation with spectacular effect is Nancy Baker Cahill's *Mushroom Cloud* project, deployed in the Pacific Ocean off of the Santa Monica Pier in February 2022 (see fig. 18). The project seeks to "recontextualize accountability" and is itself part of a larger network of both ephemeral and physically collectible objects such as videos, drawings, and NFTs. One afternoon, I decided to check out the piece for myself. I

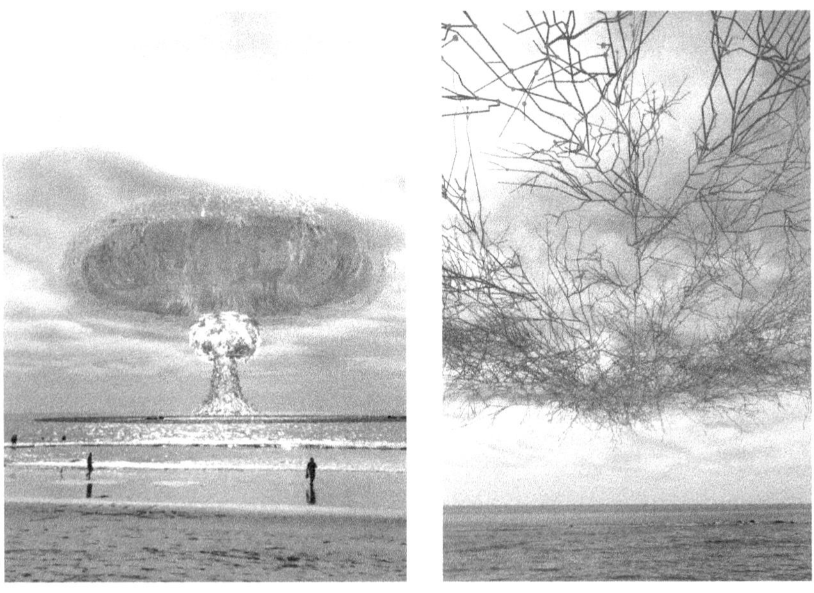

FIG. 18. *Mushroom Cloud,* Nancy Baker Cahill, 2022. (Photos taken by the author; courtesy of the artist)

hopped on the Expo Line and walked to the end of the busy Santa Monica Pier, a location I would normally avoid.

After some initial troubleshooting that required updating the 4th Wall app in which the piece is viewable, I was finally able to see the mushroom cloud eruption out in the waters, followed by the dispersive network of fibers reaching back toward the land that signal a hopeful, "connected care" model for "prioritizing equity and accountability for one another and the planet."[34] As I observed other people on the pier taking selfies with the expansive ocean as backdrop, I felt the urge to share the AR easter egg hiding just beyond their screens. What was meant as a warning about environmental disaster was taking on new resonance as Russia invaded Ukraine and readied its nuclear arsenal. The troubling specter of the mushroom cloud is the spectacular vision of the power of human destruction par excellence.

The networked tendrils did not necessarily give me hope for a new mediated communion, but they did return my attention to what was behind me: the pier, the sprawling city of Los Angeles, my own small place within a larger urban environment. And, as I walked back from the experience, I did see the pier in a transformed fashion. Whereas before I had rushed to the end to see the event, now I lingered and contemplated the various gimcracks, carnival rides, and smorgasbord of fried and sugary foods. I was drawn to a stand commemorating the Last Stop Shop at the end of Route 66, the "Mother Road," the 2,448-mile stretch starting in Chicago that largely developed the Western United States. Whereas before I had breezed past this exhibit without a second thought, it now struck me as a potent symbol of manifest destiny as well as an achingly present and material reminder of the expansive and extractive demands of industrial modernity.

I continued on and detoured through the carnival side of the pier. Families ran around giddily and lined up in nervous expectation for the adrenaline-fueled rides. The sheer glee on a child's face as a motorized airplane took off in a slow and waving circle struck me as a beautiful reminder of the importance of joy and sensation. On the one hand, perhaps I wouldn't mind so much if the Santa Monica Pier went up in a great fireball. But, on the other, I couldn't judge the escapism too harshly; it is human, necessary, and a crucial reinforcement of the social ties that are also part of Cahill's network of association and accountability.

I left the pier with much relief and headed a few hundred yards north to my usual spot on the beach. I replayed the mushroom cloud explosion a few times and was struck anew at its potency. Whereas on the pier I was pressed, surrounded by the masses, here with a few other beach goers and birds flying by, the cloud appeared less spectacular and more painful. The thought of losing a spot dear to me struck more of a chord, and the networked threads did point to a larger need for mediation and communication. Again, I felt a desire to share the cloud with those sitting around me, until my phone abruptly died, overloaded by an energy-intensive program that it was not accustomed to running. Cut off from the device, I pulled out the book I had brought along and resumed my reading.

CONCLUSION

Reality's Return

> At some point in life the world's beauty becomes enough. You don't need to photograph, paint, or even remember it. It is enough. No record of it needs to be kept and you don't need someone to share it with or tell it to. When that happens—that letting go—you let go because you can.
>
> —Toni Morrison, *Tar Baby*

> Si je ne puis pas devenir animal, je crains au moins d'être devenir bête.
>
> —James Joyce, letter to Mlle Linnossier, February 23, 1921

Geology trades in long durations. The time of complex plant and animal life began some 538.8 million years ago. *Homo sapiens* are a relative newcomer to appear on this scene. Epochs tend to last a few million years. Even the shortest measure of geologic time, an "age," normally lasts a few thousand years. Whether age, epoch, or eon, human civilization is likely to persist far into the future. But not all lifeforms and lifestyles will enjoy the same longevity. The island nation of Tuvalu may have only a few more decades. Dozens of species will not make it to next year. While much remains out of control, humans can still work toward an epoch that promotes biodiversity and environmental justice. Human empathy remains a valuable resource, especially as threats escalate into humanitarian crises. Still, it remains to be seen if this empathy will extend beyond the species to confront the growing ecological toll anthropocentrism takes on the planet.

When Saturn returns it is often a sign that one way of living has run its course. At an impasse, one must either forge a new path or risk complete annihilation through the repetition of nonviable actions. The escapist must eventually turn the power off—or, perhaps more commonly, life intervenes to expose one's limits, to force a confrontation with reality. As planetary systems continue to change, environmental disasters such as hurricanes, forest fires, and droughts increasingly affect people's lives around the world, forcing them to consider a new, détourned nature that disrupts their daily mediation. There are ways to prepare for these disasters, to insure against them, and create technological buffers, but there should also be a larger conversation on the ways in which to approach the unfolding social, emotional, and psychological damage. Change is easier said than done. "Everybody thinks of changing humanity," writes Leo Tolstoy, "and nobody thinks of changing himself."[1]

The hubris of technodeterminism promises the geoengineering of not only Earth's climate but also that of other planets. If humans sap the resources of Earth and it becomes uninhabitable, they will simply push on to the next rock down the line. The dreams of space colonies on Mars suggest an all-consuming need to go further, to see what is possible, to place future innovation over present need. While these dreams may one day come to fruition, for the time being, in the words of Mike Berners-Lee, there is no planet B. Still, the spectacular images brought to life by the Hubble telescope continue to shape wonder and amazement at the far reaches of the universe. The newly launched James Webb infrared telescope promises to deliver a window into a temporal sublime, allowing humans to look back in time to understand the early formations of stars and galaxies as well as the role of black holes in holding matter together.

In her poetry collection *Life on Mars,* Tracy K. Smith situates the miraculous promises of the interstellar age alongside both spiritual and mundane aspects of life on Earth. The poems assemble a variety of signifiers: reflections on her father's work constructing the Hubble telescope, popular culture references, the physics of dark matter, the dynamic art of David Bowie, news reports from war in Iraq, the gassing of geese at JFK Airport, pirates off the coast of Africa. Despite all of the technological, artistic, and spiritual potential, the eighth section of the titular poem enforces an ontology of earth as inescapable:

> The earth beneath us. The earth
>
> Around and above. The earth
>
> Pushing up against our houses,
>
> Complicit with gravity. The earth
>
> Ageless watching us rise and curl.
>
> Our spades, our oxen, the jagged lines

We carve into dirt. The earth
Nicked and sliced into territory.
Hacked and hollowed. Stoppered tight.
Tripwire. The earth ticking with mines,
Patient, biding its time. The earth
Floating in darkness, suspended in spin.
The earth gunning it around the sun.
The earth we ride in disbelief.
The earth we plunder like thieves.
The earth caked to mud in the belly
Of a village with no food. Burying us.
The earth coming off on our shoes.[2]

The anaphora "The earth" emphasizes both the presence of matter and a redirection of the human gaze away from virtual images toward physical reality. The earth bears witness to destructive activities with stoic endurance. The poem showcases persistence, a give and take between humans and nature.

If earth presents one absolute, time presents another. "We are all migrants through time," writes Mohsin Hamid in *Exit West*.[3] This surrealist novel presents two refugees, Nadia and Saeed, who seek to escape violence, death, and the loss of love through a series of migrations through doors that absorb all light: "Drawing close, she was struck by its darkness, its opacity, the way that it did not reveal what was on the other side, and also did not reflect what was on this side, and so felt equally like a beginning and an end."[4] The description of the door is situated alongside the other infrastructure of resources, media, and technology in the different cities to which they connect. Phones are described as antennas, divining rods that sniff out an invisible world of faraway places and imaginary zones.

Saeed and Nadia exit through a series of such doors leading to Mykonos, London, and Marin. In contrast to the technological magic of the web browser, these spaces negate present reality and thrust the pair into an entirely new environment from which they must reassemble their lives.

Passing through the door is a shocking moment of antimediation that reboots the migrant's sensorium: "It was said in those days that the passage was both like dying and like being born, and indeed Nadia experienced a kind of extinguishing as she entered the blackness and a gasping struggle as she fought to exit it, and she felt cold and bruised and damp as she lay on the floor of the room at the other side, trembling and too spent at first to stand, and she thought, while she strained to fill her lungs, that this dampness must be her own sweat."[5] The rebuilding process takes time and is navigated with the help of empathetic friends and through inventive interfaces with the urban infrastructure. In London they live in a dark halo set up around the city for migrants but disconnected from the electronic grid. Saeed works on a road dispatch while Nadia works on a crew laying spools of different pipe conduits to "connect people without requiring them to move."[6] In Marin they take advantage of wireless data, solar panels, and dew collectors to live off-grid. At one point a miniature surveillance drone crashes into their window. Hijacking these nimble technical apparatuses to survive displacement and precarity may well prove indispensable in a world marked by climate refugees and ecological disaster.

The proliferation of media and technology does not seem to be going anywhere anytime soon. This book has surveyed authors, intellectuals, artists, innovators, and institutions to gain insight into the spectacular nature of the Anthropocene. Assembling this rebel geology is a way of mediating an increasingly complex phenomenological existence on the planet. Never have questions of the natural and the artificial been more intermingled. Never has the demand for resources been more consequential. In a short text from 1977, *Forget Foucault,* Baudrillard makes an incisive aside on the ecology of energy in his discussion of reality, power, and politics: "Today especially, the real is no more than a stockpile of dead matter, dead bodies, and dead language. It still makes us feel secure today to evaluate this *stock of what is real* (let's not talk about energy: the ecological complaint hides the fact that it is not material energy which is disappearing on the species' horizon but the *energy of the real,* the reality of the real and of every serious possibility, capitalistic or revolutionary, of managing the real)."[7] A human epoch signals the stockpile of dead matter, carbon emissions, waste in all forms, species extinctions, ocean acidification, chemical saturation, and more.

With the increasing inventiveness of extractivist capitalism and the growing synthesis of technical media, Baudrillard's diagnosis largely continues to apply to the contemporary moment. Technological interests have an incentive to ramp up deliberate efforts of geoengineering to counteract adverse planetary changes. Small-scale environmental justice efforts risk losing attention to the new imperialist weight given to an epoch defining shift. The nuances of ecology encounter a bulldozer in the growing discourse of anthropogenic power.

All of these factors increase the specter of reality's dwindling energy supply. Yet this does not mean that reality has had its day. As environmental disasters increase in scope and severity, reality becomes painfully clear, demanding an ethics of action and care from the human species. Moments of antimediation provide an opportunity to release the energy of the real, a sense of liveliness that disrupts habit and passive consumption in favor of more productive forms of connection between humans and the larger biosphere. Rather than simply observing destruction on a screen, having an affective response, and moving on, antimediation encourages a longer unsettling that leads to sustained activism on environmental justice issues that do not conveniently resolve by the end of the episode. It is this energy of the real that environmental aesthetics in the twenty-first century should strive to set free.

NOTES

Introduction

1. Shawn Patterson Jr. and Michael E. Mann, "Public Disapproval of Disruptive Climate Change Protests," November 14, 2022, https://web.sas.upenn.edu/pcssm/commentary/public-disapproval-of-disruptive-climate-change-protests.
2. Paul Virilio, *The Information Bomb,* trans. Chris Turner (London: Verso, 2003), 36.
3. John Durham Peters, *The Marvelous Clouds: Toward a Philosophy of Elemental Media* (Chicago: University of Chicago Press, 2015), 3.
4. Georg Toepfer, "From *Anthropocene* to *Mediocene?* On the Use and Abuse of Stratifying the Earth's Crust by Mapping Time into Space," *ZMK* 9, no. 1 (2018): 73.
5. Don DeLillo, *White Noise* (New York: Penguin, 1985), 64.
6. DeLillo, *White Noise,* 66.
7. See, for example, Rachel A. Howell, "Investigating the Long-Term Impacts of Climate Change Communications on Individuals' Attitudes and Behavior," *Environment and Behavior* 46, no. 1 (2014): 70–101.
8. BBC Media Centre, "Watching Nature Programmes Makes You Happier New Research Reveals," 8 March 2017, https://www.bbc.co.uk/mediacentre/worldwide/2017/rhp.
9. Jonathan Crary, *Suspensions of Perception: Attention, Spectacle, and Modern Culture* (Cambridge, MA: MIT Press, 1999), 74.
10. Marc Augé, *Non-Places,* trans. John Howe (London: Verso, 2006), viii.
11. Augé, *Non-Places,* 98.
12. P. Crutzen and E. Stoermer, "The 'Anthropocene,'" *Global Change Newsletter* 41 (2000): 17–18.
13. Christophe Bonneuil and Jean-Baptiste Fressoz, *The Shock of the Anthropocene: The Earth, History, and Us* (London: Verso, 2016); Todd Braje, "Evaluating the Anthropocene: Is There Something Useful about a Geological Epoch of Humans?," *Antiquity* 90, no. 350 (2016): 504–12.
14. Astrida Neimanis, Cecilia Åsberg, and Johan Hedrén, "Four Problems, Four Directions for Environmental Humanities," *Ethics and Environment* 20, no. 1 (2015): 67–97.

15. Lesley Duxbury, "A Change in the Climate: New Interpretations and Perceptions of Climate Change through Artistic Interventions and Representation," *Weather, Climate, and Society* 2 (2010), 295.
16. Stacey Balkan, "Anthropocene and Empire," in *Think in Public: A Public Books Reader,* ed. Sharon Marcus and Caitlin Zaloom (New York: Columbia University Press, 2019), 43, 49.
17. Anders Ekström, "Exhibiting disasters: Mediation, Historicity, and Spectatorship," *Media, Culture & Society* 34, no. 4 (2012): 473.
18. T. J. Demos, *Against the Anthropocene: Visual Culture and Environment Today* (Berlin: Sternberg, 2017), 12.
19. Rob Nixon, *Slow Violence and the Environmentalism of the Poor* (Cambridge, MA: Harvard University Press, 2011), 12.
20. Caroline Levine, *The Activist Humanist: Form and Method in the Climate Crisis* (Princeton, NJ: University of Princeton Press, 2023), 12–13.
21. Oliver Wendell Holmes, *Soundings from the Atlantic* (Boston: Ticknor & Fields, 1864), 161. See also Friedrich Kittler's discussion in *Optical Media,* trans. Anthony Enns (Cambridge: Polity, 2010), 41.
22. Matthew Fuller, *Media Ecologies: Materialist Energies in Art and Technoculture* (Cambridge, MA: MIT Press: 2005), 4.
23. Anna Kornbluh, *Immediacy, or, The Style of Too-Late Capitalism* (London: Verso, 2023).
24. G. H. Chadwick, "Subdivision of Geologic Time," *Bulletin of the Geological Society of America* 41 (1930): 47–48.
25. Finis Dunaway, *Seeing Green: The Use and Abuse of Environmental Images* (Chicago: University of Chicago Press, 2015), 203.
26. Amitav Ghosh, *The Great Derangement* (Chicago: University of Chicago Press, 2016), 23.
27. Daniel Reynolds, *Media in Mind* (New York: Oxford University Press, 2019), 10.
28. Maëlle Jacqmarcq, "Environmental Activism in the Digital Age," *FLUX: International Relations Review* 11, no. 1 (2021): 41–51.
29. Jacqmarcq cites, for example, the search engine Ecosia that donates to Rainforest sustainability programs but also sells user data to encourage purchases online ("Environmental Activism," 48–49).
30. Elizabeth Swanstrom, *Animal, Vegetable, Digital* (Tuscaloosa: University of Alabama Press, 2016), 4–5.
31. Stéphane Vial, *Being and the Screen: How the Digital Changes Perception,* trans. Patsy Baudoin (Cambridge, MA: MIT Press, 2019), 52.
32. Jussi Parikka, *A Geology of Media* (Minneapolis: University of Minnesota Press, 2015); Sean Cubitt, *Finite Media: Environmental Implications of Digital Technologies* (Durham, NC: Duke University Press, 2017); Richard

Maxwell and Toby Miller, *Greening the Media* (Oxford: Oxford University Press, 2012).

33. Jussi Parikka, *The Anthrobscene* (Minneapolis: University of Minnesota Press, 2014), 5.
34. Katherine Hayles, *How We Think: Digital Media and Contemporary Technogenesis* (Chicago: University of Chicago Press, 2012), 8.
35. For a larger discussion of the remediation of "new media," see Jay David Bolter and Richard Grusin, *Remediation: Understanding New Media* (Cambridge, MA: MIT Press, 1999).
36. Nam June Paik, "Versatile Color TV Synthesizer" (1969), reprinted in *Nam June Paik: Videa 'n' Videology, 1959–1973*, ed. Judson Rosenbush (Syracuse, NY: Everson Museum of Art, 1974).
37. Alenda Chang, *Playing Nature: Ecology in Video Games* (Minneapolis: University of Minnesota Press, 2019), 65.
38. Marion Thain, "Distracted Reading: Acts of Attention in the Age of the Internet," *Digital Humanities Quarterly* 12, no. 2 (2018), http://www.digitalhumanities.org/dhq/vol/12/2/000393/000393.html (accessed June 16, 2024).

1. Aesthetics of Antimediation

1. Lorraine Daston, *Against Nature* (Cambridge, MA: MIT Press, 2019), 41.
2. Paul Winter, *Callings*, Living Music Foundation, Whole Earth Catalog Records, Stanford University Special Collections M1045, Palo Alto, CA.
3. Bernd M. Scherer, ed., *Textures of the Anthropocene: Grain, Vapor, Ray* (Cambridge, MA: MIT Press, 2015), 3.
4. Steve Mentz, *Break Up the Anthropocene* (Minneapolis: University of Minnesota Press, 2019), 9.
5. Jussi Parikka, *A Geology of Media* (Minneapolis: University of Minnesota Press, 2015), 20.
6. Sylvia Plath, *Ariel* (New York: HarperCollins, 1999), 8.
7. Andreas Weber, *Enlivenment: Toward a Poetics of the Anthropocene* (Cambridge, MA: MIT Press, 2019), 76.
8. Timothy Morton, *Humankind* (London: Verso, 2019), 188.
9. Hayles, *How We Think*, 101.
10. For more on Jerome and the Vulgate Bible, see H.A.G. Houghton, *The Latin New Testament: A Guide to Its Early History, Texts, and Manuscripts* (Oxford: Oxford University Press, 2016).
11. Robert Anderson, "Cardinal Albrecht of Brandenburg as Saint Jerome," http://ringlingdocents.org/albrechtposter.htm (accessed June 16, 2024).
12. For more on Albrecht, Luther, and Halles, see Julie Sutter, *Luther and the Cardinal* (Piccadilly: Religious Tract Society, 1883).

13. "3800 Square Foot Model of the Traveling Circus," Ringling Museum, Sarasota, Florida.
14. Christine L. Marran, *Ecology without Culture* (Minneapolis: University of Minnesota Press, 2017), 12–23.
15. James Jaehoon Lee and Joshua Beckelhimer, "Anthropocene and Empire: Discourse Networks of the Human Record," *PMLA* 135, no. 1 (2020): 127.
16. Rob Nixon, "Anthropocene 2," in *Fueling Culture: 101 Words for Energy and Environment,* ed. Imre Szeman, Jennifer Wenzel, and Patricia Yaeger (New York: Fordham University Press, 2017), 44.
17. Jamaica Kincaid, *A Small Place* (New York: Farrar, Straus & Giroux, 1988), 36–37.
18. Kathryn Yusoff, *A Billion Black Anthropocenes or None* (Minneapolis: University of Minnesota, Press, 2018), 51.
19. Teresa Brennan, *Exhausting Modernity: Grounds for a New Economy* (London: Routledge, 2000), 89.
20. Cubitt, *Finite Media,* 4.
21. Cubitt, 11.
22. McKenzie Wark, *Molecular Red: Theory for the Anthropocene* (London: Verson, 2015), 22.
23. Emmanuel Alloa, "Metaxy: Aristotle on Mediacy," in *Classics and Media Theory,* ed. Pantelis Michelakis, trans. Hayley Wood (Oxford: Oxford University Press, 2020), 148.
24. Alloa, "Metaxy," 164.
25. Alloa, 165.
26. Peters, *Marvelous Clouds,* 48; 122.
27. Tobias Menely, "Anthropocene Air," *Minnesota Review,* 83 (2014): 93; 100.
28. Alexander Galloway, Eugene Thacker, and McKenzie Wark, *Excommunication: Three Inquiries in Media and Mediation* (Chicago: University of Chicago Press, 2014), 138.
29. Kyle Bladow and Jennifer Ladino, *Toward an Affective Ecocriticism: Placing Feeling in the Anthropocene* (Lincoln: University of Nebraska Press, 2018), 3.
30. Richard P. Marzec, "Reflections on the Anthropocene Dossier," *Modern Fiction Studies* 64, no. 4 (Winter 2018): 602.
31. Gilles Deleuze, *Negotiations,* trans. Martin Joughin (New York: Columbia University Press, 1995), 127.
32. Karl Marx, *The Portable Karl Marx,* trans. Eugene Kamenka (New York: Penguin, 1983), 14.
33. Marina Vishmidt, "Mediation," in *Kunst Und Politik: Schwerpunkt: Keywords for Marxist Art History Today,* ed. Larne Abse Gogarty and Andrew Hemingway (Göttingen: V&R Unipress, 2019), 88.

34. Raymond Williams, *Keywords: A Vocabulary of Culture and Society* (New York: New Edition, 2015), 154.
35. *Oxford English Dictionary,* s.v. "Media."
36. *Oxford English Dictionary,* s.v. "Mediation."
37. Karl Marx, *Capital, Vol. 1,* trans. Ben Fowkes (New York: Vintage, 1976), 283.
38. Karl Marx, *Critique of Hegel's Philosophy of Right,* trans. Joseph O'Malley (Cambridge: Cambridge University Press, 1977), 88.
39. Robin Attfield, *Environmental Ethics: A Very Short Introduction* (Oxford: Oxford University Press, 2018), 78.
40. Henry Louis Gates Jr., *The Signifying Monkey: A Theory of African-American Literary Criticism* (Oxford: Oxford University Press, 2014), 61.
41. Eugene Thacker and Yves Citton, "Antimediation," *Multitudes* 51, no. 4 (2012): 100; emphasis in original.
42. Galloway, Thacker, Wark, *Excommunication,* 59–60.
43. Galloway, Thacker, Wark, 63.
44. J. M. Synge, *Collected Works: Volume 2,* ed. Alan Price (London: Oxford University Press, 1966), 8.
45. Deleuze, *Negotiations,* 122.
46. Kornbluh, *Immediacy,* 111.
47. Deborah Vankin, "Why Is an AR Mushroom Cloud Exploding over the Santa Monica Pier?," *LA Times,* February 11, 2022, https://www.latimes.com/entertainment-arts/story/2022-02-11/ar-mushroom-cloud-over-the-santa-monica-pier-frieze-week.
48. Augé, *Non-Places,* x.
49. Elizabeth DeLoughery, *Allegories of the Anthropocene* (Durham, NC: Duke University Press, 2019), 5–6.
50. William Shakespeare, *The Tempest* (New York: Penguin, 1970), 33.
51. "Stage History," Royal Shakespeare Company, accessed June 16, 2024, https://www.rsc.org.uk/the-tempest/about-the-play/stage-history.
52. Stephen Orgel, *The Tempest* (Oxford: Oxford University Press, 1987), 56–62.
53. Ursula Heise, *Sense of Place and Sense of Planet: The Environmental Imagination of the Global* (Oxford: Oxford University Press, 2008); Kelly Oliver, *Earth and World: Philosophy after the Apollo Missions* (New York: Columbia University Press, 2015); Tiago de Luca, *Planetary Cinema: Film, Media, and the Earth* (Amsterdam: Amsterdam University Press, 2021).
54. De Luca, *Planetary Cinema,* 40.
55. Lewis Fry Richardson, *Weather Prediction by Numerical Process* (Cambridge: Cambridge University Press, 2007), 219.
56. Joanna Page, "Planetary Art beyond the Human: Rethinking Agency in the Anthropocene," *The Anthropocene Review* 7, no. 3 (2000): 273.

57. Edwards, *A Vast Machine: Computer Models, Climate Data, and the Politics of Global Warming* (Cambridge, MA: MIT Press, 2010), 25.
58. Edwards, *Vast Machine*, 94.
59. Edwards, 96.

2. Natural Historicities

1. Christian Nyampeta, *Sometimes It Was Beautiful*, organized by Xiaoyu Weng, Guggenheim Museum, New York City, April 2021.
2. Galloway, Thacker, Wark, *Excommunication*, 54; emphasis in original.
3. Bonneuil and Fressoz, *Shock of the Anthropocene*, 117.
4. Leo Marx, "Technology: The Emergence of a Hazardous Concept," *Social Research* 64, no. 3 (Fall 1997): 970.
5. Oscar Wilde, *Intentions* (Portland, ME: T. B. Mosher, 1904), 3.
6. Charles Dickens, *Hard Times* (London: Bradbury & Evans, 1854), 338.
7. Wilde, *Intentions*, 30.
8. *The Germ*, ed. William Michael Rosetti (London: Elliot Stock, 1901), 31.
9. Christina Rossetti, *The Complete Poems of Christina Rosetti* (New York: Penguin, 2001), 796.
10. Wilde, *Intentions*, 36.
11. Thomas Hood, *The Serious Poems* (London: E. Moxon, 1868), 131.
12. Pliny the Elder, *Natural History*, trans. Philemon Holland (London: George Barclay, 1847), 18.
13. John F. Healy, "Introduction," in *Natural History* by Pliny the Elder (New York: Penguin, 2004), 12.
14. Francis Bacon, *The Works of Francis Bacon* (Philadelphia: M. Murphy, 1876), 124.
15. Charles Lyell, *Principles of Geology* (London: John Murray, 1875), 8.
16. Lyell, *Principles of Geology*, 76.
17. Baruch Spinoza, *Ethics, including the Improvement of the Understanding*, trans. R. H. M. Elwes (New York: Prometheus, 1989), 12–13.
18. "The Imperial Institute," *Bulletin of the Imperial Institute, Vol. 1* (London: Jas. Truscott & Son, 1903), ii.
19. *British Empire Exhibition Handbook* (Plaistow: Curwen, 1924), ZLIB 15/50/31, National Archives, Kew, 10.
20. *British Empire Exhibition Handbook*, 21
21. "Pageant of Empire," British Empire Exhibition internal memo, September 25, 1923, ZLIB 15/50/31, National Archives, Kew.
22. "Pageant of Empire," 1.
23. *The Pageant of Empire Programme, Part 3* (London: Fleetway, 1924), 8.

24. Virginia Woolf, "Thunder at Wembley," in *A Room of One's Own and Other Essays* (London: Folio Society, 2000), 168.
25. Woolf, "Thunder at Wembley," 169.
26. "The Empire Story-Land," Imperial Institute Literature, CO 1045/236, National Archives, Kew, 2.
27. H. A. F. Lindsay, *The Empire Film Library Catalogue,* 1939, Imperial Institute Literature, CO 1045/236, National Archives, Kew, 1939.
28. H. A. F. Lindsay, "India's Place in Empire Films" (1939), Imperial Institute Literature, CO 1045/236, National Archives, Kew, 5.
29. H. A. F. Lindsay, "The Imperial Institute and its Work for the Empire," *Analyst* 66, no. 783 (June 1941): 222–27.
30. "The Imperial Institute in War-Time," *Times Educational Supplement,* February 15, 1941.
31. Letter from G. P. Humphreys-Davies, January 11, 1946, Imperial Institute Literature, CO 1045/236, National Archives, Kew.
32. Lindsay, "India's Place in Empire Films," 8.
33. "Exhibition Galleries," *1953 Annual Report,* Imperial Institute Literature, CO 1045/236, National Archives, Kew.
34. "The Imperial Institute and the Imperial College," *House of Lords Official Report* 179.14, December 9, 1952, Imperial Institute Literature, CO 1045/236, National Archives, Kew, 851.
35. Francis Bacon, *New Atlantis* (Oxford: Clarendon, 1915), 35.
36. Hannah Arendt, *Between Past and Future* (New York: Penguin, 1954), 58–60.
37. Dipesh Chakrabarty, "The Climate of History: Four Theses," *Critical Inquiry* 35, no. 2 (Winter 2009): 206.
38. Susan Buck-Morss, *The Dialectics of Seeing: Walter Benjamin and the Arcades Project* (Cambridge, MA: MIT Press, 1989), 59.
39. Buck-Morss, *Dialectics of Seeing,* 65.
40. César Aira, *An Episode in the Life of a Landscape Painter,* trans. Chris Andrew (New York: New Directions, 2006), 77–78.
41. Chimamanda Ngozi Adichie, "The Headstrong Historian," *New Yorker,* June 16, 2008, www.newyorker.com/magazine/2008/06/23/the-headstrong-historian.
42. Adichie, "Headstrong Historian."
43. Adichie.
44. Shaobo Xie, "Is the World Decentered?," in *Global Fissures: Postcolonial Fusions,* ed. Clara A. B. Joseph and Janet Wilson, Cross/Cultures 85 (Amsterdam: Rodopi, 2006), 74.
45. Omar El Akkad, "Faster than We Thought: What Stories Will Survive Climate Change?," Literary Hub, September 16, 2019, https://lithub.com/faster-than-we-thought-what-stories-will-survive-climate-change/.

3. From Global Village to Anthropocene

1. Harold Innis, *Empire and Communications* (Lanham, MD: Rowman & Littlefield, 2007), 13.
2. Innis, *Empire and Communications,* 190–97.
3. James Nisbet, *Ecologies, Environments, and Energy Systems in Art of the 1960s and 1970s* (Cambridge: MIT Press, 2014), 153.
4. Innis, *Empire and Communications,* 196.
5. Pierre Teilhard De Chardin, *The Phenomenon of Man,* trans. Bernard Wall (New York: Harper Perennial, 2008), 182.
6. De Chardin, *Phenomenon of Man,* 214.
7. Kevin Kelley, *What Technology Wants* (New York: Penguin, 2011), 8.
8. Aristotle, *Politics,* trans. C. D. C. Reeve (Indianapolis: Hackett, 1998), 216.
9. William Connolly, *Facing the Planetary: Entangled Humanism and the Politics of Swarming* (Durham, NC: Duke University Press, 2017), 9.
10. Don DeLillo, *Point Omega* (New York: Scribner, 2010), 52–53.
11. DeLillo, *Point Omega,* 5–6.
12. Tom Wolfe, "Foreword," in Marshall McLuhan, *Understanding Me: Lectures and Interviews,* ed. Stephanie McLuhan and David Staines (Cambridge, MA: MIT Press, 2005), xvii.
13. Wolfe, "Foreword," 37.
14. Wolfe, 49.
15. Wolfe, 67.
16. Jakob von Uexküll, *A Foray into the Worlds of Animals and Humans,* trans. Joseph D. O'Neil (Minneapolis: University of Minnesota Press, 2010), 41.
17. Marshall McLuhan, *Understanding Media: The Extensions of Man* (New York: McGraw Hill, 1965), 43.
18. Anaïs Nin, *Under a Glass Bell and Other Stories* (Chicago: Swallow, 1948), 23.
19. McLuhan, *Understanding Media,* 130.
20. Rachel Carson, *Silent Spring* (New York: Houghton Mifflin, 2002), 203.
21. David Harvey, *The Condition of Postmodernity* (Oxford: Blackwell, 1990), 240.
22. John Paul Sartre, *Nausea,* trans. Lloyd Alexander (New York: New Directions, 1964), 10–11.
23. Walker Percy, *The Moviegoer* (New York: Knopf, 1962), 41.
24. Percy, *Moviegoer,* 166.
25. Percy, 16.
26. Jennifer Fay, *Inhospitable World: Cinema in the Time of the Anthropocene* (New York: Oxford University Press, 2018), 4.
27. Jerzy Kosinski, *Being There* (New York: Harcourt Brace, 1970), 6.
28. Kosinski, *Being There,* 63.

29. Ovid, *Metamorphoses,* trans. Rolfe Humphries (Bloomington: Indiana University Press, 1955), 70.
30. Ovid, *Metamorphoses,* 71.
31. Percy, *Moviegoer,* 196.
32. George Maciunus, *Fluxus Manifesto,* Festus Fluxorum Fluxus, Düsseldorf, 1963.
33. "Nam June Paik: Moon Is the Oldest TV," *American Masters,* PBS, May 16, 2023.
34. John Cage, *John Cage: Writer,* ed. Richard Kostelanetz (New York: Limelight Editions, 1993), 155.
35. *Whole Earth Field Guide,* ed. Caroline Maniaque-Benton (Cambridge, MA: MIT Press, 2016), 5.
36. *Whole Earth Field Guide,* 40.
37. Hanns Reich, *The World from Above* (New York: Hill & Wang, 1966), 11.
38. Nisbet, *Ecologies, Environments, and Energy Systems,* 80.
39. Nisbet, 120.
40. Aldo Leopold, *The Sand County Almanac* (New York: Oxford University Press, 2020), 192.
41. Steven T. Byington, "On Interference with the Environment," *New Freewoman* 1, no. 7 (September 1913): 123.
42. Edward T. Hall, *The Hidden Dimension* (Garden City, NY: Doubleday, 1966), 50–55.
43. Norbert Wiener, *Cybernetics* (Cambridge, MA: MIT Press, 1961), 22.
44. William Wordsworth, *Poems* (New York: Knopf, 1995), 14.
45. Marshall McLuhan, *Culture Is Our Business* (New York: Ballantine, 1970), 180.
46. Beryl Korot and Phyllis Gershuny, "Editorial," *Radical Software* 1, no. 11 (Spring 1970): 1.

4. From Ecological *Détournement* to Rebel Geology

1. London Psychogeophysics Summit, "What is Psychogeophysics?," Mute, August 4, 2010, https://www.metamute.org/community/your-posts/what-psychogeophysics.
2. Jules Verne, *Works of Jules Verne, Volume 13* (New York. F. Tyler Daniels, 1911), 413.
3. Andrew Hussey, "Spectacle, Simulation, and Spectre: Debord, Baudrillard and the ghost of Marx," *Parallax* 7, no. 3 (2001): 68.
4. Jean Baudrillard, "Requiem for the Media," in *Utopie: Texts and Projects, 1967–1978,* ed. Craig Buckley and Jean-Lous Violeau (Los Angeles: Semiotext(e), 2011), 265.

5. Guy Debord, *The Society of the Spectacle,* trans. Donald Nicholson-Smith (Brooklyn, NY: Zone, 1994), 12.
6. McLuhan, *Understanding Media,* 67.
7. Guy Debord, "Theses on Traffic," in *Situationist International Anthology,* ed. Ken Knabb (Berkeley, CA: Bureau of Public Secrets, 2006), 69.
8. Chinua Achebe, "Benin Road," in *Beware, Soul-Brother, and Other Poems* (London: Heinemann 1972), 6.
9. Chinua Achebe, "Benin Road," *Collected Poems* (New York: Penguin, 2004), 4.
10. W. H. Auden, *Selected Poems* (New York: Vintage, 1979), 207.
11. Debord, *Society of the Spectacle,* 19.
12. Debord, 92.
13. Debord, 118.
14. Greil Marcus, "The Long Walk of the Situationist International," in *Guy Debord and the Situationist International,* ed. Tom McDonough (Cambridge: MIT Press, 2002), 7.
15. Wark, *Molecular Red,* 16.
16. Jason Moore, ed., *Anthropocene or Capitalocene? Nature, History, and the Crisis of Capitalism* (Oakland, CA: Kairos/PM, 2016), 85.
17. Moore, *Anthropocene or Capitalocene?,* 114. Moore also notes the emergence of the digital in the contemporary era, citing the role of satellite imagery for planetary monitoring, but he does not take into account the ways in which capitalism also profits from the information age.
18. Naomi Klein, *This Changes Everything* (New York: Simon & Schuster, 2014), 158.
19. Manfredo Tafuri, *Architecture and Utopia: Design and Capitalist Development* (Cambridge: MIT Press, 1976), 4.
20. Tafuri, *Architecture and Utopia,* 7.
21. McDonough, *Guy Debord and the Situationist International,* xi.
22. Ian Baucom, "'Moving Centers': Climate Change, Critical Method, and the Historical Novel," *Modern Language Quarterly* 76, no. 2 (June 2015): 156.
23. Patrick Modiano, *In the Café of Lost Youth,* trans. Chris Clarke (New York: New York Review, 2016), 15.
24. A reference to Ed van der Elsken's *Love on the Left Bank,* published in 1956.
25. McDonough, *Guy Debord and the Situationist International,* xi.
26. Tafuri, *Architecture and Utopia,* 73.
27. McDonough, *Guy Debord and the Situationist International,* xii.
28. Modiano, *In the Café of Lost Youth,* 51.
29. Modiano, 40.

30. Modiano, 53.
31. Modiano, 76.
32. Modiano, 77.
33. Notably, Guy de Vere is the name of the fiancé in Edgar Allen Poe's poem "Lenore," which similarly laments the death of a young woman.
34. Modiano, *In the Café of Lost Youth,* 108.
35. Established in 1783 by Louis XVI. Now Mines ParisTech, one of the most prominent engineering schools in France with a strong international reputation.
36. The Mexico was closed in 1995. The space is now Corcoran's Irish Pub.
37. Modiano, *In the Café of Lost Youth,* 49.
38. Modiano, 60.
39. Modiano, 139.
40. Timothy Morton, *Dark Ecology: For a Logic of Future Coexistence* (New York: Columbia University Press, 2016), 130.
41. Alain Badiou, *Philosophy and the Event,* trans. Louise Burchill (Cambridge, MA: Polity, 2013), 34–35.
42. Knabb, *Situationist International Anthology,* 203; emphasis in original.
43. "Dialectical Utopia," trans. Jean-Marie Clarke, in *Utopie,* 62.
44. "Dialectical Utopia," 62.
45. René Lourau, "Notes on Marxism and Aesthetics," trans. Jean-Marie Clarke, in *Utopie,* 49.
46. Jean-Paul Jungmann, "Paper Cities," trans. Jean-Marie Clarke, in *Utopie,* 100.
47. Jean Aubert, "Becoming Outdated," trans. Jean-Marie Clarke, in *Utopie,* 88.
48. Jean Baudrillard, "The Ephemeral," trans. Jean-Marie Clarke, in *Utopie,* 77.
49. "The Mystique of Environment," trans. Jean-Marie Clarke, in *Utopie,* 255.
50. "Mystique of the Environment," 257.
51. Jean Baudrillard, "Requiem for the Media," 266.

5. Cyber Saturnalia

1. Sherry Turkle, *Alone Together: Why We Expect More from Technology and Less from Each Other* (New York: Basic Books, 2011), xlviii.
2. Jacques Derrida, "Structure, Sign, and Play in Human Sciences," in *Writing and Difference,* trans. Alan Bass (Chicago: University of Chicago Press, 1978).
3. Gilbert Simondon, "Technical Mentality," in *Gilbert Simondon: Being and Technology,* ed. Arne De Boever, Alex Murray, Job Roffe, and Ashley Woodward (Edinburgh: Edinburgh University Press, 2012), 9.
4. David Packard, "Publishing Scholarly Compilations by Computer," *Computers and the Humanities* 4, no. 1 (1969): 76.

5. Packard, "Publishing Scholarly Compilations," 76.
6. Hayles, *How We Think,* 10.
7. Hayles, 81.
8. Paul Virilio, *The Information Bomb,* 18.
9. Walter Benjamin, *Illuminations* (New York: Mariner, 2019), 187.
10. Chang, *Playing Nature,* 23.
11. For more information, see "Playing for the Planet," Green Game Jam, accessed August 6, 2024, https://www.playing4theplanet.org/green-game-jam.
12. Vial, *Being and the Screen,* 52.
13. Ted Chiang, *Exhalation* (New York: Knopf, 2019), 109.
14. Chiang, *Exhalation,* 163.
15. Bernard Stiegler, *Technics and Time, 3: Cinematic Time and the Question of Malaise,* trans. Stephen Barker (Palo Alto, CA: Stanford University Press, 2011), 2.
16. Gilbert Simondon, "Technical Mentality," trans. Arne de Boever, in *Gilbert Simondon: Being and Technology,* ed. Arne De Boever, Alex Murray, Jon Roffe, and Ashley Woodward (Edinburgh: Edinburgh University Press, 2012), 9.
17. Yves Citton, *The Ecology of Attention,* trans. Barnaby Norman (Cambridge: Polity, 2017), 4.
18. Citton, *Ecology of Attention,* 22.
19. Citton, *Ecology of Attention,* 104.
20. Dominic Pettibon, *Infinite Distraction* (Cambridge: Polity, 2018), 11.
21. Pettibon, *Infinite Distraction,* 29.
22. Pettibon, 95.
23. Wilde, *Intentions,* 167.
24. Jean Baudrillard, *The Ecstasy of Communication,* trans. Bernard Schütze and Caroline Schütze (South Pasadena, CA: Semiotext(e), 2012), 26.
25. Macrobius, *Saturnalia,* trans. Percival Vaughan Davies (New York: Columbia University Press, 1969), 27.
26. Bernard Stiegler, *Technics and Time, 2: Disorientation,* trans. Stephan Barker (Palo Alto, CA: Stanford University Press, 2008), 3.
27. Jonathan Crary, *24/7: Late Capitalism and the Ends of Sleep* (London: Verso, 2013).
28. Patricia Lockwood, *No One Is Talking About This* (London: Bloomsbury, 2021), 12–13.
29. Lockwood, *No One Is Talking,* 30.
30. Lockwood, 167.
31. Lockwood, 205.
32. Jenny Offill, *Weather* (New York: Vintage, 2021), 39.
33. Offill, *Weather,* 51.

34. Most of these works can be accessed at the Electronic Literature Organization directory: (https://directory.eliterature.org).
35. In correspondence with the artist.
36. Hayles, *How We Think,* 176.

6. Digital Twins

1. Simon Kofe, "Minister Kofe's Video Statement: COP26," YouTube video, 4:04, November 8. 2021, https://www.youtube.com/watch?v=hpPTFGwFExg.
2. Simon Kofe, "Rising Sea Levels Force Tuvalu to Move to the Metaverse: COP27 Speech," YouTube video, 3:36, November 15, 2022, https://www.youtube.com/watch?v=lXpeO5BgAOM.
3. Cage, *John Cage,* 111.
4. Cage, 112.
5. Tung-Hui Hu, *A Prehistory of the Cloud* (Cambridge, MA: MIT Press, 2015), 66.
6. Hu, *Prehistory of the Cloud,* 7.
7. J. R. Carpenter, *The Gathering Cloud,* http://luckysoap.com/thegatheringcloud (accessed June 16, 2024).
8. Carpenter, *Gathering Cloud,* no. 5.
9. Jon W. Sparks, "New Software Review Is Stimulating Winner," Whole Earth Catalogue Collection, M1045, Stanford University Special Collections, Palo Alto, CA.
10. David Bunnell quoted in Susan Chace, "Publishers Rush to Software Books, Trying to Cash In on Computers," *Wall Street Journal,* May 18, 1983.
11. *Whole Earth Review* (Winter 1983): 5, Whole Earth Catalogue Collection, M1045, Stanford University Special Collections, Palo Alto, CA.
12. *Whole Earth Review* (Winter 83): 5.
13. John Markoff, "Whole Earth State-of-Art Rapping," *New York Times,* August 15, 1989.
14. Stewart Brand, "Your TV Set of the Future," *TV Guide,* June 4, 1988.
15. Hu, *Prehistory of the Cloud,* x.
16. Stewart Brand, *The Media Lab: Inventing the Future at MIT.* (New York: Penguin, 1987), 259.
17. E. M. Forster, *The Collected Tales of E. M. Forster* (New York: The Modern Library, 1959), 145.
18. Forster, *Collected Tales,* 154.
19. Forster, 153.
20. Office of the US Surgeon General, *Our Epidemic of Loneliness and Isolation,* 2023, https://www.hhs.gov/sites/default/files/surgeon-general-social-connection-advisory.pdf (accessed June 16, 2024).

21. Simone Peinkofer, "The Dark Side of Amazon Returns," *Fortune,* June 14, 2023, https://fortune.com/2023/06/14/amazon-returns-ecommerce-how-bad-big-problem-816-billion/.
22. Stiegler, *Technics and Time, 3,* 38.
23. S. J. Ahn, Elise Ogle, Kristine L. Nowak, Kara T. McGillicuddy, and Jeremy N. Bailenson, "Experiencing Nature: Embodying Animals in Immersive Virtual Environments Increases Inclusion of Nature in Self and Involvement with Nature," *Journal of Computer-Mediated Communication* 21, no. 6 (November 2016): 405.
24. This project has been extended and enriched as *The Crystal Reef,* a virtual-reality film project by Cody Karutz, Jeremy Bailenson, and Lauren Knapp, that debuted in 2016 at the Tribeca Film Festival.
25. Ahn, "Experiencing Nature," 410.
26. Patricia Hernandez, "Things Get Weird When You Look at Pokémon as an Allegory for Slavery," Kotaku, February 11, 2013, https://kotaku.com/things-get-weird-when-you-look-at-pokemon-as-an-allegor-5983399.
27. Leejiah J. Dorward, John C. Mittermeier, Chris Sandbrook, and Fiona Spooner, "Pokémon Go: Benefits, Costs, and Lessons for the Conservation Movement," *Conservation Letters* 10, no. 1 (2017): 160–65.
28. Nora N. Khan, "Conceptual Thinking," in *For Submersion* (Los Angeles: Clockshop, 2023), 9.
29. Michael Specter, "The Climate Fixers," *New Yorker,* May 7, 2012, https://www.newyorker.com/magazine/2012/05/14/the-climate-fixers.
30. Emma Strubell, Ananya Ganesh, and Andrew McCallum, "Energy and Policy Considerations for Deep Learning in NLP," in *Proceedings of the 57th Annual Meeting of the Association for Computational Linguistics,* July 2019, https://aclanthology.org/volumes/P19-1 (accessed June 16, 2024).
31. Josh Dzieza, "Inside the AI Factory," *New York Magazine,* June 20, 2023. https://nymag.com/intelligencer/article/ai-artificial-intelligence-humans-technology-business-factory.html.
32. Refik Anadol, "Glacier Dreams: at the Intersection of Art, Science & Technology," interview on Designboom, March 11, 2023, https://www.designboom.com/art/interview-refik-anadol-multisensory-generative-installation-glacier-dreams-art-dubai-03-11-2023.
33. Ricardo Lopez, "The Common Collective versus Personal Specificity: AI's Failure to Capture Nuance." Seminar Paper for ENG 490: Climate Fiction, Spring Semester, 2023, CSUDH.
34. Vankin, "Why Is an AR Mushroom Cloud Exploding?"

Conclusion

1. Leo Tolstoy, *Pamphlets,* trans. Aylmer Maude (Christchurch: Free Age, 1900), 29.
2. Tracy K. Smith, *Life on Mars* (Minneapolis: Graywolf, 2011), 40–41.
3. Mohsin Hamid, *Exit West* (New York: Riverhead, 2017), 209.
4. Hamid, *Exit West,* 103.
5. Hamid, 104.
6. Hamid, 181.
7. Jean Baudrillard, *Forget Foucault,* trans. Phil Beitchman, Nicole Dufresne, Lee Hildreth and Mark Polizzotti (Cambridge, MA: MIT Press, 1998), 46.

BIBLIOGRAPHY

Achebe, Chinua. *Beware, Soul-Brother, and Other Poems.* London: Heinemann, 1972.

———. *Collected Poems.* New York: Penguin, 2004.

Adichie, Chimamanda Ngozi. "The Headstrong Historian." *New Yorker,* June 16, 2008.

Ahn, S. J., Joshua Bostick, Elise Ogle, Kristine L. Nowak, Kara T. McGillicuddy, Jeremy N. Bailenson. "Experiencing Nature: Embodying Animals in Immersive Virtual Environments Increases Inclusion of Nature in Self and Involvement with Nature." *Journal of Computer-Mediated Communication* 21, no. 6 (November 2016): 399–419.

Aira, César. *An Episode in the Life of a Landscape Painter.* Translated by Chris Andrews. New York: New Directions, 2006.

Alloa, Emmanuel. "Metaxy: Aristotle on Mediacy." Translated by Hayley Wood. In *Classics and Media Theory,* edited by Pantelis Michelakis. Oxford: Oxford University Press, 2020. DOI: 10.1093/oso/9780198846024.003.0006 (accessed July 10, 2024).

Anadol, Refik. "Glacier Dreams: at the Intersection of Art, Science & Technology." Designboom, March 11, 2023. https://www.designboom.com/art/interview-refik-anadol-multisensory-generative-installation-glacier-dreams-art-dubai-03-11-2023/.

Arendt, Hannah. *Between Past and Future.* New York: Penguin, 1954.

Aristotle. *Politics.* Translated by C. D. C. Reeve. Indianapolis: Hackett, 1998.

Aronson, M. J. "Review of Edouard Le Roy." *Journal of Philosophy* 27, no. 18 (1930): 502–3.

Attfield, Robin. *Environmental Ethics: A Very Short Introduction.* Oxford: Oxford University Press, 2018.

Augé, Marc. *Non-Places.* Translated by John Howe. London: Verso, 2006.

Bacon, Francis. *New Atlantis.* Oxford: Clarendon, 1915.

———. *The Works of Francis Bacon.* Philadelphia: M. Murphy, 1876.

Badiou, Alain. *Philosophy and the Event.* Translated by Louise Burchill. Cambridge, MA: Polity, 2013.

Balkan, Stacey. "Anthropocene and Empire." *Think in Public: A Public Books Reader,* edited by Sharon Marcus and Caitlin Zaloom, 41–52. New York: Columbia University Press, 2019.

Baucom, Ian. "'Moving Centers': Climate Change, Critical Method, and the Historical Novel." *Modern Language Quarterly* 76, no. 2 (June 2015): 137–57.

Baudrillard, Jean. *The Ecstasy of Communication.* Translated by Bernard Schütze and Caroline Schütze. South Pasadena, CA: Semiotext(e), 2012.

———. *Forget Foucault.* Translated by Phil Beitchman, Nicole Dufresne, Lee Hildreth, and Mark Polizzotti. Cambridge, MA: MIT Press, 1998.

BBC Media Centre. "Watching Nature Programmes Makes You Happier New Research Reveals." March 8, 2017. https://www.bbc.co.uk/mediacentre/worldwide/2017/rhp.

Benjamin, Walter. *Illuminations.* Translated by Harry Zohn. New York: Mariner, 2019.

Bladow, Kyle, and Jennifer Ladino. *Toward an Affective Ecocriticism: Placing Feeling in the Anthropocene.* Lincoln: University of Nebraska Press, 2018.

Bolter, Jay David, and Richard Grusin. *Remediation: Understanding New Media.* Cambridge, MA: MIT Press, 1999.

Bonneuil, Christophe, and Jean-Baptiste Fressoz. *The Shock of the Anthropocene: The Earth, History, and Us.* London: Verso, 2016.

Braje, Todd. "Evaluating the Anthropocene: Is There Something Useful about a Geological Epoch of Humans?" *Antiquity* 90, no. 350 (2016): 504–12.

Brand, Stewart. *The Media Lab: Inventing the Future at MIT.* New York: Penguin, 1987.

Brennan, Teresa. *Exhausting Modernity: Grounds for a New Economy.* London: Routledge, 2000.

British Empire Exhibition Handbook. Plaistow: Curwen, 1924.

Buck-Morss, Susan. *The Dialectics of Seeing: Walter Benjamin and the Arcades Project.* Cambridge, MA: MIT Press, 1989.

Bulletin of the Imperial Institute, Vol. 1. London: Jas. Truscott & Son, 1903.

Byington, Steven T. "On Interference with the Environment." *New Freewoman* 1, no. 7 (September 1913): 121–23.

Cage, John. *John Cage: Writer.* Edited by Richard Kostelanetz. New York: Limelight Editions, 1993.

Carson, Rachel. *Silent Spring.* New York: Houghton Mifflin, 2002.

Chadwick, G. H. "Subdivision of Geologic Time." *Bulletin of the Geological Society of America* 41 (1930): 47–48.

Chakrabarty, Dipesh. "The Climate of History: Four Theses." *Critical Inquiry* 35, no. 2 (Winter 2009): 197–222.

Chang, Alenda. *Playing Nature: Ecology in Video Games.* Minneapolis: University of Minnesota Press, 2019.

Chiang, Ted. *Exhalation.* New York: Knopf, 2019.

Citton, Yves. *The Ecology of Attention.* Translated by Barnaby Norman. Cambridge: Polity, 2017.

Connolly, William. *Facing the Planetary: Entangled Humanism and the Politics of Swarming.* Durham, NC: Duke University Press, 2017.

Crary, Jonathan. *24/7: Late Capitalism and the Ends of Sleep.* London: Verso, 2013.

———. *Suspensions of Perception: Attention, Spectacle, and Modern Culture.* Cambridge, MA: MIT Press, 1999.

Crutzen, Paul, and E. Stoermer. "The 'Anthropocene.'" *Global Change Newsletter* 41 (2000): 17–18.

Cubitt, Sean. *Finite Media: Environmental Implications of Digital Technologies.* Durham, NC: Duke University Press, 2017.

Daston, Lorraine. *Against Nature.* Cambridge, MA: MIT Press, 2019.

De Chardin, Pierre Teilhard. *The Phenomenon of Man.* Translated by Bernard Wall. New York: Harper Perennial, 2008.

Debord, Guy. *Commentaries on the Society of the Spectacle.* Translated by Malcolm Imrie. London: Verso, 1988.

———. *The Society of the Spectacle.* Translated by Donald Nicholson-Smith. Brooklyn, NY: Zone, 1994.

Deleuze, Gilles. *Negotiations.* Translated by Martin Joughin. New York: Columbia University Press, 1995.

DeLillo, Don. *Point Omega.* New York: Scribner, 2010.

———. *White Noise.* New York: Penguin, 1985.

DeLoughery, Elizabeth. *Allegories of the Anthropocene.* Durham, NC: Duke University Press, 2019.

Demos, T. J. *Against the Anthropocene: Visual Culture and Environment Today.* Berlin: Sternberg, 2017.

Derrida, Jacques. "Structure, Sign, and Play in Human Sciences." *Writing and Difference.* Translated by Alan Bass. Chicago: University of Chicago Press, 1978.

Dorward, Leejiah J., John C. Mittermeier, Chris Sandbrook, and Fiona Spooner. "Pokémon Go: Benefits, Costs, and Lessons for the Conservation Movement." *Conservation Letters* 10, no. 1 (2017): 160–65.

Dickens, Charles. *Hard Times.* London: Bradbury & Evans, 1854.

Dunaway, Finis. *Seeing Green: The Use and Abuse of Environmental Images.* Chicago: University of Chicago Press, 2015.

Duxbury, Lesley. "A Change in the Climate: New Interpretations and Perceptions of Climate Change through Artistic Interventions and Representation." *Weather, Climate, and Society* 2 (2010): 294–99.

Edwards, Paul. *A Vast Machine: Computer Models, Climate Data, and the Politics of Global Warming.* Cambridge, MA: MIT Press, 2010.

Ekström, Anders. "Exhibiting Disasters: Mediation, Historicity, and Spectatorship." *Media, Culture & Society* 34, no. 4 (2012): 472–87.

Fay, Jennifer. *Inhospitable World: Cinema in the Time of the Anthropocene.* New York: Oxford University Press, 2018.

Forster, E. M. *The Collected Tales of E. M. Forster.* New York: Modern Library, 1959.

Fuller, Matthew. *Media Ecologies: Materialist Energies in Art and Technoculture.* Cambridge, MA: MIT Press, 2005.

Galloway, Alexander, Eugene Thacker, and McKenzie Wark. *Excommunication: Three Inquiries in Media and Mediation.* Chicago: University of Chicago Press, 2014.

Gates, Henry Louis, Jr. *The Signifying Monkey: A Theory of African-American Literary Criticism.* Oxford: Oxford University Press, 2014.

Ghosh, Amitav. *The Great Derangement.* Chicago: University of Chicago Press, 2016.

Hall, Edward T. *The Hidden Dimension.* Garden City, NY: Doubleday, 1966.

Hamid, Mohsin. *Exit West.* New York: Riverhead, 2017.

Haraway, Donna. *Staying with the Trouble: Making Kin in the Cthulucene.* Durham, NC: Duke University Press, 2016.

Harvey, David. *The Condition of Postmodernity.* Oxford: Blackwell, 1990.

Hayles, Katherine. *How We Think: Digital Media and Contemporary Technogenesis.* Chicago: University of Chicago Press, 2012.

Holmes, Oliver Wendell. *Soundings from the Atlantic.* Boston: Ticknor & Fields, 1864.

Hood, Thomas. *The Serious Poems.* London: E. Moxon, 1868.

Houghton, H. A. G. *The Latin New Testament: A Guide to Its Early History, Texts, and Manuscripts.* Oxford: Oxford University Press, 2016.

Howell, Rachel A. "Investigating the Long-Term Impacts of Climate Change Communications on Individuals' Attitudes and Behavior." *Environment and Behavior* 46, no. 1 (2014): 70–101.

Hu, Tung-Hui. *A Prehistory of the Cloud.* Cambridge, MA: MIT Press, 2015.

Hussey, Andrew. "Spectacle, Simulation, and Spectre: Debord, Baudrillard, and the Ghost of Marx." *Parallax* 7, no. 3 (2001): 63–72.

"Imperial Institute Bill of 1924." Imperial Institute: Literature. CO 1045/236. National Archives, Kew, UK.

Innis, Harold. *Empire and Communications.* Lanham, MD: Rowman & Littlefield, 2007.

Iovino, Serenella, and Serpil Opperman, eds. *Material Ecocriticism.* Bloomington: Indiana University Press, 2014.

Jacqmarcq, Maëlle. "Environmental Activism in the Digital Age." *FLUX: International Relations Review* 11, no. 1 (2021): 41–51.

Kelley, Kevin. *What Technology Wants.* New York: Penguin, 2011.

Kittler, Friedrich. *Optical Media.* Translated by Anthony Enns. Cambridge: Polity, 2010.

Kincaid, Jamaica. *A Small Place.* New York: Farrar, Straus & Giroux, 1988.
Klein, Naomi. *This Changes Everything.* New York: Simon & Schuster, 2014.
Knabb, Ken, ed. *Situationist International Anthology.* Berkeley, CA: Bureau of Public Secrets, 2006.
Kornbluh, Anna. *Immediacy, or, The Style of Too-Late Capitalism.* London: Verso, 2023.
Korot, Beryl, and Phyllis Gershuny. "Address to Readers." *Radical Software* 1, no. 1 (Spring 1970): n.p.
Kosinski, Jerzy. *Being There.* New York: Harcourt Brace, 1970.
Lee, James Jaehoon, and Joshua Beckelhimer. "Anthropocene and Empire: Discourse Networks of the Human Record." *PMLA* 135, no.1 (2020): 110–29.
Leopold, Aldo. *The Sand County Almanac.* New York: Oxford University Press, 2020.
Levine, Caroline. *The Activist Humanist: Form and Method in the Climate Crisis.* Princeton, NJ: University of Princeton Press, 2023.
Lindsay, H. A. F. *The Empire Film Library Catalogue.* 1939. Imperial Institute: Literature. CO 1045/236. National Archives, Kew, UK.
———. "The Imperial Institute and its Work for the Empire." *Analyst* 66, no. 783 (June 1941): 222–25.
———. "India's Place in Empire Films." 1939. Imperial Institute: Literature. CO 1045/236. National Archives, Kew, UK.
Lockwood, Patricia. *No One Is Talking About This.* London: Bloomsbury, 2021.
London Psychogeophysics Summit. "What Is Psychogeophysics?" *Mute,* August 4, 2010. https://www.metamute.org/community/your-posts/what-psychogeophysics.
Lyell, Charles. *Principles of Geology.* London: John Murray, 1875.
Macrobius. *Saturnalia.* Translated by Percival Vaughan Davies. New York: Columbia University Press, 1969.
Marcuse, Herbert. *Hegel's Ontology and the Theory of Historicity.* Translated by Seyla Benhabib. Cambridge, MA: The MIT Press, 1987.
Marran, Christine L. *Ecology without Culture.* Minneapolis: University of Minnesota Press, 2017.
Marzec, Richard P. "Reflections on the Anthropocene Dossier." *Modern Fiction Studies* 64, no. 4 (Winter 2018): 585–616.
Marx, Karl. *Capital, Vol. 1.* Translated by Ben Fowkes. New York: Vintage, 1976.
———. *Critique of Hegel's Philosophy of Right.* Translated by Joseph O'Malley. Cambridge: Cambridge University Press, 1977.
———. *The Portable Karl Marx.* Translated by Eugene Kamenka. New York: Penguin, 1983.
Marx, Leo. "Technology: The Emergence of a Hazardous Concept." *Social Research* 64, no. 3 (Fall 1997): 965–88.

Maxwell, Richard, and Toby Miller. *Greening the Media*. Oxford: Oxford University Press, 2012.

McDonough, Tom, ed. *Guy Debord and the Situationist International*. Cambridge, MA: MIT Press, 2002.

McLuhan, Marshall. *Culture Is Our Business*. New York: Ballantine, 1970.

———. *Understanding Me: Lectures and Interviews*. Edited by Stephanie McLuhan and David Staines. Cambridge, MA: MIT Press, 2005.

———. *Understanding Media: The Extensions of Man*. New York: McGraw Hill, 1965.

Menely, Tobias. "Anthropocene Air." *Minnesota Review*, 83 (2014): 93–101.

Mentz, Steve. *Break Up the Anthropocene*. Minneapolis: University of Minnesota Press, 2019.

Modiano, Patrick. *In the Café of Lost Youth*. Translated by Chris Clarke. New York: New York Review, 2016.

Moore, Jason, ed. *Anthropocene or Capitalocene? Nature, History, and the Crisis of Capitalism*. Oakland, CA: Kairos/PM, 2016.

Morton, Timothy. *Dark Ecology: For a Logic of Future Coexistence*. New York: Columbia University Press, 2016.

———. *Humankind: Solidarity with Non-Human People*. London: Verso, 2017.

"Nam June Paik: Moon Is the Oldest TV." *American Masters*, PBS, May 16, 2023.

Neimanis, Astrida, Cecilia Åsberg, and Johan Hedrén. "Four Problems, Four Directions for Environmental Humanities." *Ethics and Environment* 20, no. 1 (2015): 67–97.

Nin, Anaïs. *Under a Glass Bell and Other Stories*. Chicago: Swallow, 1948.

Nisbet, James. *Ecologies, Environments, and Energy Systems in Art of the 1960s and 1970s*. Cambridge: MIT Press, 2014.

Nixon, Rob. "Anthropocene 2." In *Fueling Culture: 101 Words for Energy and Environment*, edited by Imre Szeman, Jennifer Wenzel, and Patricia Yaeger, 43–46. New York: Fordham University Press, 2017.

———. *Slow Violence and the Environmentalism of the Poor*. Cambridge, MA: Harvard University Press, 2011.

Offill, Jenny. *Weather*. New York: Vintage, 2021.

Ovid. *Metamorphoses*. Translated by Rolfe Humphries. Bloomington: Indiana University Press, 1955.

Packard, David. "Publishing Scholarly Compilations by Computer." *Computers and the Humanities* 4, no. 1 (1969): 75–80.

Page, Joanna. "Planetary Art beyond the Human: Rethinking Agency in the Anthropocene." *Anthropocene Review* 7, no. 1 (2000): 273–94.

Pageant of Empire Programme, Part 3. London: Fleetway, 1924.

Parikka, Jussi. *The Anthrobscene*. Minneapolis: University of Minnesota Press, 2014.

———. *A Geology of Media.* Minneapolis: University of Minnesota Press, 2015.
Percy, Walker. *The Moviegoer.* New York: Knopf, 1962.
Peters, John Durham. *The Marvelous Clouds: Toward a Philosophy of Elemental Media.* Chicago: University of Chicago Press, 2015.
Pettibon, Dominic. *Infinite Distraction.* Cambridge: Polity, 2018.
Plath, Sylvia. *Ariel.* New York: HarperCollins, 1999.
Pliny the Elder. *Natural History.* Translated by Philemon Holland. London: George Barclay, 1847.
Reynolds, Daniel. *Media in Mind.* New York: Oxford University Press, 2019.
Reich, Hanns. *The World from Above.* New York: Hill & Wang, 1966.
Richardson, Lewis Fry. *Weather Prediction by Numerical Process.* Cambridge: Cambridge University Press, 2007.
Rosetti, Christina. *The Complete Poems of Christina Rosetti.* New York: Penguin, 2001.
Rosetti, William Michael, ed. *The Germ.* London: Elliot Stock, 1901.
Sartre, John Paul. *Nausea.* Translated by Lloyd Alexander. New York: New Directions, 1964.
Scherer, Bernd M., ed. *Textures of the Anthropocene: Grain, Vapor, Ray.* Cambridge, MA: MIT Press, 2015.
Schön, Donald A. *Technology and Change: The New Heraclitus.* New York: Delacorte, 1967.
Shakespeare, William. *The Tempest.* New York: Penguin, 1970.
Simondon, Gilbert. "Technical Mentality." In *Gilbert Simondon: Being and Technology,* edited by Arne De Boever, Alex Murray, Jon Roffe, and Ashley Woodward, 1–16. Translated by Arne De Boever. Edinburgh: Edinburgh University Press, 2012.
Smith, Tracy K. *Life on Mars.* Minneapolis, MN: Graywolf, 2011.
Spinoza, Baruch. *Ethics, including the Improvement of the Understanding.* Translated by R. H. M. Elwes. New York: Prometheus, 1989.
Spivak, Gayatri Chakravorty. *A Critique of Postcolonial Reason: Toward a History of the Vanishing Present.* Cambridge, MA: Harvard University Press, 1999.
Stiegler, Bernard. *Technics and Time, 2: Disorientation.* Translated by Stephan Barker. Palo Alto, CA: Stanford University Press, 2008.
———. *Technics and Time, 3: Cinematic Time and the Question of Malaise.* Translated by Stephen Barker. Palo Alto, CA: Stanford University Press, 2011.
Stone, Caroline. "The Sustainability of Our Digital Environments: The Language of the Upgrade Path and E-waste." *Green Letters: Studies in Ecocriticism* 18, no. 3 (2014): 286–96.
Strubell, Emma, Ananya Ganesh, Andrew McCallum. "Energy and Policy Considerations for Deep Learning in NLP." In *Proceedings of the 57th*

Annual Meeting of the Association for Computational Linguistics. July 2019. arXiv:1906.02243.
Sutter, Julie. *Luther and the Cardinal.* Piccadilly: Religious Tract Society, 1883.
Swanstrom, Elizabeth. *Animal, Vegetable, Digital.* Tuscaloosa: University of Alabama Press, 2016.
Synge, J. M. *Collected Works: Volume 2,* edited by Alan Price. London: Oxford University Press, 1966.
Tafuri, Manfredo. *Architecture and Utopia: Design and Capitalist Development.* Cambridge, MA: MIT Press, 1976.
Thacker, Eugene, and Yves Citton. "Antimediation." *Multitudes* 51, no. 4 (2012): 99–110.
Thain, Marion. "Distracted Reading: Acts of Attention in the Age of the Internet." *Digital Humanities Quarterly* 12, no. 2 (2018). http://www.digitalhumanities.org/dhq/vol/12/2/000393/000393.html.
Toepfer, Georg. "From *Anthropocene* to *Mediocene?* On the Use and Abuse of Stratifying the Earth's Crust by Mapping Time into Space." *ZMK* 9, no. 1 (2018): 73–84.
Tolstoy, Leo. *Pamphlets.* Translated by Aylmer Maude. Christchurch: Free Age, 1900.
Turkle, Sherry. *Alone Together: Why We Expect More from Technology and Less from Each Other.* New York: Basic Books, 2011.
Uexküll, Jakob von. *A Foray into the Worlds of Animals and Humans.* Translated by Joseph D. O'Neil. Minneapolis: University of Minnesota Press, 2010.
Utopie: Texts and Projects, 1967–1978. Edited by Craig Buckley and Jean-Louis Violeau. Translated by Jean-Marie Clarke. Los Angeles: Semiotext(e), 2011.
Verne, Jules. *Works of Jules Verne, Volume 13.* Edited by Charles F. Horne. New York: F. Tyler Daniels, 1911.
Vial, Stéphane. *Being and the Screen: How the Digital Changes Perception.* Translated by Patsy Baudoin. Cambridge, MA: MIT Press, 2019.
Vishmidt, Marina. "Mediation." In *Keywords for Marxist Art History Today,* edited by Larne Abse Gogarty and Andrew Hemingway, 87–96. Göttingen: V&R Unipress, 2019.
Virilio, Paul. *The Information Bomb.* Translated by Chris Turner. London: Verso, 2003.
Wark, McKenzie. *Molecular Red: Theory for the Anthropocene.* London: Verso, 2015.
Weber, Andreas. *Enlivenment: Toward a Poetics of the Anthropocene.* Cambridge, MA: MIT Press, 2019.
Whole Earth Catalog Records. M1045. Stanford University Special Collections. Palo Alto, CA.
Whole Earth Field Guide. Edited by Caroline Maniaque-Benton. Cambridge, MA: MIT Press, 2016.

Wiener, Norbert. *Cybernetics.* Cambridge, MA: MIT Press, 1961.
Wilde, Oscar. *Intentions.* Portland, ME: T. B. Mosher, 1904.
Williams, Raymond. *Keywords: A Vocabulary of Culture and Society.* New York: New Edition, 2015.
Willmott, Glenn. *McLuhan, or Modernism in Reverse.* Toronto: University of Toronto Press, 1996.
Woolf, Virginia. "Thunder at Wembley." *A Room of One's Own and Other Essays.* London: Folio Society, 2000.
Wordsworth, William. *Poems.* New York: Knopf, 1995.
Xie, Shaobo. "Is the World Decentered?" In *Global Fissures: Postcolonial Fusions,* edited by Clara A. B. Joseph and Janet Wilson, 53–78. Cross/Cultures, 85. Amsterdam: Rodopi, 2006.
Yusoff, Kathryn. *A Billion Black Anthropocenes or None.* Minneapolis: University of Minnesota, Press, 2018.

INDEX

abstraction, 10, 13, 23–26, 90–92
Achebe, Chinua, 68, 101–2
activism: and aesthetics, 5–10, 36; and digital media, 11, 99, 134–37, 157–59, 162; and materialism, 28–29; and spectacle, 1, 113
Adams, Ansel, 7
Adichie, Chimamanda Ngozi, 16, 67–68
Adorno, Theodor, 33, 64–65, 151
aesthetics: and activism, 6–11, 21–23, 98, 113; of cinema, 84, 111–12; and ecology, 74, 127–28; of environment, 52, 106, 108, 114, 160; of holism, 15, 89–93; of media, 78, 154; and nature, 47–49; of networks, 14, 158–59; and sensation, 66–67; and urbanism, 102–5
affect, 6, 13, 23, 30–32, 91, 106, 127–28, 137, 169
Aira, César, 16, 65–66
algorithm, 9, 25, 45, 112, 125, 133, 156
alienation, 5, 11, 79–80, 99, 104, 110–11, 127–29, 150
Amazon, 150
Anadol, Refik, 159–60
Anglocene, 47, 52
anthrodecentrism, 29–30, 32
Anthropocene: and activism, 22, 29, 112–13, 116, 127; and allegory, 39; and empire, 47; and film, 84; and geoengineering, 104–5; and the great acceleration, 63; and material reality, 9, 22–25, 53; and media theory, 6, 31, 53, 168
Anthropocene Project (Haus der Kulturen der Welt), 23
anti-environment, 17, 36, 74, 78, 95
antimediation: and activism, 2–3, 142; and digital media, 121–22, 127, 132, 157–61; and self-awareness, 86–87, 168; theories of, 36–42
Apple, 122, 141, 151
À Rebours (Huysmans), 52
Arendt, Hannah, 64
Aristotle, 30–31, 77, 128
artificial intelligence (AI), 124, 143–44, 148, 158–62
attention, 4–7, 10–15, 36, 47, 125–28, 133–34
Aubert, Jean, 114–15, 117
Auden, W. H., 102
Augé, Marc, *Non-Places*, 5, 39, 106
augmented reality (AR), 39, 133, 143, 154–58
autofiction, 14, 134–37
automobile, 76, 100–102, 125, 145
autopoiesis, 14, 32, 43, 77, 82, 87, 143

Bacon, Francis, 53–55, 64, 110
Badiou, Alain, 112
Balog, James, 38, 159
Bataille, Georges, 16, 36–37
Baudrillard, Jean, 13, 99–100; *The Ecstasy of Communication*, 128–29; "The Ephemeral," 114–16; *Forget Foucault*, 168; "Requiem for the Media," 116
Being There (Kosinski), 17, 85–88
"Benin Road" (Achebe), 101–2
Benjamin, Walter, 37, 65, 123
Blake, William, 79–80
Blue Marble, The, 42
Brand, Stewart, 90–96, 145–48
Brennan, Teresa, 30
British Empire, 16, 56–63, 67, 73–75, 89
British Empire Exhibition, 57–59
Byington, Steven, 92–93

Cage, John, 89, 143
Cahill, Nancy Baker, 39, 162–63
Callings (Winter), 22
capitalism, 8–9, 24, 29–30, 57, 65, 87–89, 101–13, 155, 168–69
Capitalocene, 103
Caravaggio, 87–88
carbon offsets, 2, 28, 97
Cardinal Albrecht of Brandenburg as Saint Jerome, 26–28
Carpenter, J. R., 137, 144–45
Carson, Rachel, 36, 74, 82
catharsis, 128
Chadwick, George Halcott, 8
Chakrabarty, Dipesh, 48, 64, 68
Chasing Ice, 38, 159
ChatGPT, 160–61
Chiang, Ted, 18, 124–25, 140, 155
circus, 28–29, 37, 49
Citton, Yves, 36–37, 126–27
Clean Air Act of 1956, 8–9
climate change, 12–13; and antimediation, 38–39; and carbon offsets, 28; denial of, 98–99; and geoengineering, 102–5; and sea level rise, 142–43; and spectacle, 84; and technology, 112–13, 134–35, 155, 159–61
Climate Change Conference of the Parties (COP), 142
climate refugees, 143, 167–68
cloud computing, 18, 142–45, 148
colonialism, 9, 29–30, 46–47, 58–73, 89, 92, 100, 154–55
Committee for the Investigation of Atmospheric Pollution, 8
comparative media, 4, 13–15, 45, 74, 133, 137
Conrad, Joseph, 98, 101
conservation, 75, 97, 123, 137, 155, 160
Constant (artist), 114–15
Cranach the Elder, Lucas, 26–28
Crary, Jonathan, 5, 13, 132
Critique of Hegel's Philosophy of Right (Marx), 33–34
Cybernetics (Wiener), 93–94
Darwin, Charles, 37–38, 49, 54, 56
data, 8, 37, 43–45, 61–66, 86, 94, 106, 130, 133, 141–44, 150, 155–60
Debord, Guy, 1, 5, 13, 17, 36, 99–106, 111, 128
decolonization, 17, 46, 73, 101
Deepwater Horizon (2010), 21
defamiliarization, 38
Deleuze, Gilles, 32, 37–38
DeLillo, Don, 15; *Point Omega,* 77–78; *White Noise,* 4
dérive, 84, 100, 155
Derrida, Jacques, 122
détournement, 36–38, 100–103, 109–13, 116
Dickens, Charles, 48–49, 59, 66
digital detox, 18, 141
digital twin, 2, 18, 125, 144, 162
distraction, 6, 15, 32, 127
Dragged Mass (Heizer), 17, 91–92
Dump VR (Leischner and Peterkova), 153–54

earth music, 22
Earthrise, 42, 91
Earthrise Media, 1, 158
earthworks, 74, 91
El Akkad, Omar, 68–69
electronic literature, 13, 18, 133, 137–41
Eliasson, Olafur, 117
emissions, 2, 9, 13, 28, 53–54, 75, 103–5, 126, 143, 150, 158
Empire and Communications (Innis), 73–75
energy, 9, 28, 30, 37, 50, 74, 81–82, 121, 128–29, 158, 168–69
environmental aesthetics, 14, 21–25, 48–52, 74–75, 99, 104–13, 128–29, 154, 160, 169
environmental humanities, 2, 5, 7, 14–15
environmental justice, 6, 15, 21, 32, 92–93, 105, 116, 127–29
Episode in the Life of a Landscape Painter, An (Aira), 65–66

ethnic environmentalism, 29, 68
e-waste, 2, 9, 126
Exhalation (Chiang), 18, 124–25
Exit West (Hamid), 167–68
extinction, 5, 9, 39, 44, 93, 97, 124, 140, 161
extraction, 2, 4, 8–9, 46, 54, 63, 73, 103, 168

film, 13–16, 38–39, 45–47, 59–63, 77, 83–90, 111
formalism, 7, 38, 82, 100
Forster, E. M., 18, 148–50
For Submersion (Rosalena), 155–58
Foucault, Michel, 128, 168

Galloway, Alexander, 37, 47
Garden (Roberts), 139–40
Gates, Henry Louis, Jr., 16, 36
geoengineering, 28, 64, 76, 81, 105, 109, 137, 158, 166, 168
geology, 13, 54–56, 156, 165
Germ, The, 50
Ghosh, Amitav, 10, 84
"Global Groove" (Paik), 17, 88–89
globalization, 15, 42, 54, 68–69, 75, 81, 103, 112, 123
global warming, 12, 75
Google, 12, 141
Goya, Francisco, 130–31
Great Smog of 1952, 8
Green Game Jam, 2, 123
greenhouse gasses, 75. *See also* emissions
green media, 2, 6, 144
greenwashing, 22, 97, 160

Hamid, Mohsin, 167–68
Haraway, Donna, 13, 22, 33
Hard Times (Dickens), 49–50
Hayles, Katherine, 13, 25, 122–23, 141
"Headstrong Historian, The" (Adichie), 67–68
Heizer, Michael, 17, 91–92
historicity, 58, 64–69, 145
holism, 15, 33, 36, 42, 55, 75, 90–93, 110, 125
Hollywood, 4, 113, 148
Holmes, Oliver Wendell, 7
Holocene, 32
Homo sapiens, 5, 8, 29, 129, 165
"Houseboat" (Nin), 80–82
Hubble telescope, 166
Huxley, Julian, 75
Huysmans, Joris Karl, 52
hylomorphic, 77

IBM, 12
immediacy, 8, 38–39
immersion, 5, 10–13, 33, 45, 63, 78, 84, 128, 147, 151, 159
Imperial Institute, 16, 56–63, 73–75, 89, 155
impressionism, 10, 50–52
Inconvenient Truth, An, 4
Innis, Harold, 73–75
interface, 15, 107, 122–23, 128, 149
International Union for the Conservation of Nature, 75
In the Café of Lost Youth (Modiano), 17, 105–12
isolation, 5, 18, 45, 134, 149–50

Joyce, James, 134, 165
Jungmann, Jean-Paul, 114
Just Stop Oil, 1

Kelly, Kevin, 76–77
Kincaid, Jamaica, 29–30
Kittler, Friedrich, 7, 13
Klein, Naomi, 103–4
Kofe, Simon, 142–43
Kornbluh, Anna, 7–8, 38
Kosinski, Jerzy, 17, 85–88

"Lady Lazarus" (Plath), 24, 82
Latour, Bruno, 14
Leischner, Vojtech, 153–54

Leopold, Aldo, 91–92
Levine, Caroline, 7
Life on Mars (Smith), 1, 166–67
Lindsay, H. A. F., 59–62
lithium, 23
Lockwood, Patricia, 18, 134–37
Lyell, Charles, 53–56

"Machine Stops, The" (Forster), 18, 148–50
Macrobius, 130–33
Marx, Karl, 16, 32–36, 121
Marx, Leo, 47–48
Marxism, 16, 32–36, 99, 102, 107
materialism, 14, 28, 30, 32, 83, 91
materiality, 11, 14, 16, 23, 29, 35, 87, 99, 144
May 1968 protests, 98–100, 113–16
McLuhan, Marshall, 13, 17, 36, 74, 78–82, 85, 95, 99–100
Mediacene, 3–12, 47, 102, 126
media ecology, 7, 10, 18, 39, 92, 116, 126–27, 141, 160, 168
medial, 3, 14–15, 31, 78
mediasphere, 126–27
media theory, 6, 11–14, 17, 30–32, 37, 42, 68–69, 100, 113–14
mediation, 3, 6–8, 12–18; aesthetic, 22, 66; digital, 45, 121, 124; ecological, 30; judicial, 21; Marxist, 32–34; material, 30, 33; ontological, 30–31, 33; passive, 100; political, 33, 77, 99; technological, 112
mediator, 3, 22, 32–34, 93, 126, 161
metafiction, 14
metaverse, 11, 151
meteorology, 13, 16, 42–44, 65, 144
Microsoft, 12, 122
Midsummer Night's Dream, A, 34–35, 39
MIT Media Lab, 148
Modiano, Patrick, 17, 105–12
mondialisation, 15, 39, 102–5
Monet, Claude, 50–52
Morrison, Tony, 165
Morton, Timothy, 14, 25, 112
Moviegoer, The (Percy), 17, 83–84, 87
Mushroom Cloud (Cahill), 39, 162–63
music, 5, 14, 22, 37, 57–58, 88, 140, 143, 148

Narcissus, 86–88, 94–95
natural history, 16, 47–48, 53–57, 64–69, 73, 102–3, 130, 155
natural sciences, 16, 52–56, 102
nature, 3, 33, 39–43, 48–60, 64–66, 101–10, 122–30, 143, 152
nature cure, 83
Nausea (Sartre), 82–83
new materialism, 14, 23–25, 28–32, 83
new media, 10–11, 14, 104, 121–30, 137–38, 145, 150–51
Newton (Blake), 79–80
Nietzsche, Friedrich, 21
NIMBY, 21
Nin, Anaïs, 80–82
Nisbet, James, 74, 90–91
Nixon, Rob, 6, 9, 29
nonfungible token (NFT), 160, 162
nonhuman, 12, 18, 23–25, 31, 37, 98, 106, 141, 152, 161
No One Is Talking About This (Lockwood), 134–36
noosphere, 17, 75–78, 126
nuclear, 25, 64, 163
Nyampeta, Christian, 16, 46–47

object-oriented ontology, 14, 83, 128
observation, 8–10, 43, 53, 123
Offill, Jenny, 18, 136–37
On the Origin of Species (Darwin), 37–38, 56
Ovid, 86, 129

Paik, Nam June, 14, 17, 88–89
Parikka, Jussi, 13, 23
participation, 11–12, 22, 152–55
Percy, Walker, 17, 83–84, 87
Peterkova, Natalia, 153–54

Peters, John Durham, 3, 31
Pettibon, Dominic, 127
Phanerozoic, 8
planetary media, 11–16, 31–32, 42–45, 90–91, 117–18, 161
Planet Earth (TV show), 4
Plath, Sylvia, 24, 82
Pliny the Elder, 53–54, 130
poioumena, 14, 39–42, 94
Pokémon Go, 154–55
Polli, Andrea, 117
praxis, 11, 90, 100, 113–14, 130
pre-Raphaelites, 49–50, 52
Principles of Geology (Lyell), 55–56
psychogeography, 17, 84, 98–100, 107, 155
psychogeophysics, 98–99
Purchase of the North Pole, The / Topsy Turvy (Verne), 98–99

race, 29–30, 60, 68, 76, 90, 113
Radical Software, 95
realism, 10, 48–49, 125
reality, 8–13, 23–25, 49, 125, 148, 153, 165–69
rebel geology, 13, 98, 168
representation, 4–11, 39–40, 57–58, 66, 90, 94, 114, 159
revolution, 11, 16–17, 58, 69, 88, 97–98
Richardson, Lewis Fry, 42–45
Ringling, John, 27–29
Roberts, Owen, 139–40
Roosegaarde, Daan, 117
Rosalena, Sarah, 155–58
Rossetti, Christina, 50

Sand County Almanac (Leopold), 91–92
Santa Barbara Oil Spill (1969), 21
Sartre, Jean Paul, 82–83, 87
Saturnalia, 18, 37, 86, 129–33
"Seedlings_:From Humus," 138–39
Shakespeare, 3, 34–35, 40–42, 148
Silent Spring (Carson), 36, 82
Silicon Valley, 17, 89
Simondon, Gilbert, 17, 121–22, 126
Situationist International, 17, 98–99, 102–6, 114
slavery, 29–30, 47, 154
slow violence, 6, 36
Small Place, A (Kincaid), 29–30
smart phone, 2, 23, 99, 107
Smith, Tracy K., 1, 166–67
Society of the Spectacle (Debord), 5, 100–102, 104–5
Sometimes It Was Beautiful (Nyampeta), 46–47
Songs of the Humpback Whale (Payne), 22
spectacle, 1–12, 28, 39–42, 100–105, 111–13, 127–29, 151–53
Spinoza, Baruch, 55–56
Spivak, Gayatri Chakravorty, 68
Stiegler, Bernard, 126, 131, 151
sublime, 6, 40–42, 55–56, 108, 149, 166
Synge, J. M., 37–38

Tafuri, Manfredo, 104–17, 128
technogenesis, 124–25
technology, 7–14, 25, 42–45, 74–82, 112–18, 121–25, 132–41, 143–62
technosphere, 13, 53, 74, 97
Teilhard de Chardin, Pierre, 75–78
television, 4, 13–14, 83–89, 98–99, 126, 138
Tempest, The, 3, 40–42
Thacker, Eugene, 13, 16, 31, 36–37
"Thunder at Wembley" (Woolf), 58–59
Topsy Turvy / The Purchase of the North Pole (Verne), 98–99
transmedial, 14–15
"Tree" (VR), 151–52
Turkle, Sherry, 12, 121
Turrell, James, 117
Tuvalu, 142–44
24 Hour Psycho (Gordon), 77–78

Uexküll, Jakob von, 80
Umwelt, 80

Understanding Media: The Extensions of Man (McLuhan), 80–82
UNESCO, 12, 75
United Kingdom, 47, 56–63
urbanism, 80, 99–104, 114
Utopie, 17, 98–100, 114–18

Verne, Jules, 98–99
Vial, Stéphane, 12, 123–24
video art, 77–78, 88–89
video games, 14, 123, 154–55
Virilio, Paul, 2, 123
Virtual Human Interaction Lab (Stanford), 152–53
virtual reality (VR), 25, 121, 124–26, 142–44, 151–54

Wark, McKenzie, 30, 103
Waterloo Bridge (Monet), 50–52
Watts rebellion (1965), 113
Weather (Offill), 136–37
Weather Prediction by Numerical Process (Richardson), 42–44
Whole Earth Catalog, 76, 89–96
Whole Earth 'Lectronic Link (WELL), 146–47
Whole Earth Software Catalog, 145–46
Wiener, Norbert, 90, 93–94
Wilde, Oscar, 48–53, 66–67, 95, 115, 128
Williams, Raymond, 33
Winter, Paul, 22
Woolf, Virginia, 58–59, 134
Wordsworth, William, 94–95
World Economic Forum, 12, 160
Wuerker, Matthew, 146–47

YouTube, 132–33
Yusoff, Kathryn, 30

Cultural Frames, Framing Culture

Changed Men: Veterans in American Popular Culture after World War II
Erin Lee Mock

We, Us, and Them: Affect and American Nonfiction from Vietnam to Trump
Douglas Dowland

Criminal Cities: The Postcolonial Novel and Cathartic Crime
Molly Slavin

Skimpy Coverage: Sports Illustrated *and the Shaping of the Female Athlete*
Bonnie M. Hagerman

Story Revolutions: Collective Narratives from the Enlightenment to the Digital Age
Helga Lenart-Cheng

Institutional Character: Collectivity, Agency, and the Modernist Novel
Robert Higney

Walk the Barrio: The Streets of Twenty-First-Century Transnational Latinx Literature
Cristina Rodriguez

Fashioning Character: Style, Performance, and Identity in Contemporary American Literature
Lauren S. Cardon

Neoliberal Nonfictions: The Documentary Aesthetic from Joan Didion to Jay-Z
Daniel Worden

Dandyism: Forming Fiction from Modernism to the Present
Len Gutkin

Terrible Beauty: The Violent Aesthetic and Twentieth-Century Literature
Marian Eide

Women Writers of the Beat Era: Autobiography and Intertextuality
Mary Paniccia Carden

Stranger America: A Narrative Ethics of Exclusion
Josh Toth

Fashion and Fiction: Self-Transformation in Twentieth-Century American Literature
Lauren S. Cardon

American Road Narratives: Reimagining Mobility in Literature and Film
Ann Brigham

The Arresting Eye: Race and the Anxiety of Detection
Jinny Huh

Failed Frontiersmen: White Men and Myth in the Post-Sixties American Historical Romance
James J. Donahue

Composing Cultures: Modernism, American Literary Studies, and the Problem of Culture
Eric Aronoff

Quirks of the Quantum: Postmodernism and Contemporary American Fiction
Samuel Chase Coale

Chick Lit and Postfeminism
Stephanie Harzewski

American Iconographic: "National Geographic," Global Culture, and the Visual Imagination
Stephanie L. Hawkins

Wanted: The Outlaw in American Visual Culture
Rachel Hall

Male Armor: The Soldier-Hero in Contemporary American Culture
Jon Robert Adams

African Americans and the Culture of Pain
Debra Walker King

Against the Unspeakable: Complicity, the Holocaust, and Slavery in America
Naomi Mandel

I'm No Angel: The Blonde in Fiction and Film
Ellen Tremper

Visions of the Maid: Joan of Arc in American Film and Culture
Robin Blaetz

Writing War in the Twentieth Century
Margot Norris

The Golden Avant-Garde: Idolatry, Commercialism, and Art
Raphael Sassower and Louis Cicotello

Kodak and the Lens of Nostalgia
Nancy Martha West

www.ingramcontent.com/pod-product-compliance
Ingram Content Group UK Ltd.
Pitfield, Milton Keynes, MK11 3LW, UK
UKHW042330090226
467866UK00005B/15